Low-Volume Road Engineering

Design, Construction, and Maintenance

Low-Volume Road Engineering

Design, Construction, and Maintenance

Robert A. Douglas

Golder Associates, Canada

CRC Press
Taylor & Francis Group
Boca Raton London New York

CRC Press is an imprint of the
Taylor & Francis Group, an **informa** business
A SPON PRESS BOOK

CRC Press
Taylor & Francis Group
6000 Broken Sound Parkway NW, Suite 300
Boca Raton, FL 33487-2742

First issued in paperback 2017

ISBN-13: 978-1-4822-1263-1 (hbk)
ISBN-13: 978-1-138-74815-6 (pbk)

Visit the Taylor & Francis Web site at
http://www.taylorandfrancis.com

and the CRC Press Web site at
http://www.crcpress.com

To Professor Lynne Irwin, mentor and friend, and Ted Douglas, who inspired in his three children a love of the written word.

Contents

Preface

In 1983, as a brand-new assistant professor on probation, I inherited a university course in transportation engineering applied to industrial forestry. I was appalled; the course was terrible. Little of it was relevant to the transportation of logs and wood chips to mills and processing plants. It even included airport planning theory. Really.

I spent the summer revamping the course, setting it up to cover such things as road location, geometric design, haul truck performance, scheduling, and haul road maintenance management. At the same time, I was tweaking another course in the application of geotechnical engineering to forest haul road design. I had a great deal of help from the venerable Ross Silversides and the well-known Thom Bjerkelund, two of the brilliant people I've been fortunate to meet.

The material for the forest transportation course was cobbled together from many, many disparate sources. I scoured the civil engineering, mechanical engineering, and forestry literature for scraps relevant to industrial forest operations. I handed out tons of notes to students, incidentally doing my small bit to bolster the Canadian forest industry.

Eventually, I realized this was a massively inefficient way to provide students with the material they needed. Why not write a book? Wouldn't it just be a matter of changing all those bullet point lists into full sentences?

The result—a couple of years later—was *Delivery: The Transportation of Raw Natural Resource Product from Roadside to Mill.* It was self-published at the university and provided to my students and those of similar forest engineering programs elsewhere, and a handful of other members of industry who came to know about it.

Meantime, I published as much research as I could, but I soon found resource road research was a bit of an orphan. I had to disguise my road engineering stuff as "forest operations" material in order to get it accepted by conveners of forestry conferences and symposia and generalize my forestry-focused work in order to get it accepted by editors of mainstream civil and geotechnical engineering journals.

Then I had the good fortune to attend the 8th International Conference on Low-Volume Roads (LVR) in Reno, Nevada. At the conference, it became clear that I had been dabbling in the formal field of LVR, and here was a community of like-minded professionals working at a high level on these roads, quite unapologetically. I met such notables as Asif Faiz, Lynne Irwin, and Alex Visser. These people conducted research, hosted international conferences and national symposia, and participated actively in the vast committee system of the U.S. Transportation Research Board. It was like the opening of a door I didn't even know existed.

In countless discussions, it became clear to me that there are actually two types of LVR, distinguished by the vehicles that use them. On the one hand, there are LVR that cater to light traffic—the traffic volume is low, and the vehicles are light. Rural roads, regional roads, farm-to-market roads, roads in developing countries, roads in parks, roads on First Nations lands, and roads in the far north all fit into this category. Most of the LVR literature falls into this domain. On the other hand, there are the LVR carrying what can be enormously heavy haul trucks. Again, the traffic volumes are low, but the vehicles using the roads can weigh as much as 600 tons. Haul roads for surface mining, oil sands extraction, oil and gas development, and industrial forestry all fit into this category.

I now had a good conceptual framework for LVR and had become a member of a strong professional community devoted to LVR. But still it seemed students and practising professionals were stuck with drawing the literature from "here, there, and everywhere," as the song goes. A single reference that summarized the material needed for practice, and that pointed to sources of follow-up information, was needed.

Based on the success of *Delivery*, with the unfailing encouragement of Lynne Irwin at Cornell University, I set about writing such a book.

Low-Volume Road Engineering: Design, Construction, and Maintenance draws together relevant material based on my experience in Canada, the United States, the United Kingdom, New Zealand, and West Africa. Reference is made to as much of the English language material as possible, citing specifications and practices in the United States, Canada, the United Kingdom, New Zealand, Australia, and South Africa. It presents the material as a standard engineering reference, rather than a handbook or manual, and it's comfortable with metric and U.S. customary units.

I do hope you find the book as interesting and enjoyable as I found writing it and that it becomes your "go-to" reference for LVR engineering.

Robert A. Douglas

Acknowledgments

It takes the support of many willing, enthusiastic individuals to complete a book project.

Lynne Irwin, father of the U.S. LTAP (Local Technical Assistance Program) system, was there at the beginning of the project many years ago, and with it to completion, with a much-needed gentle prod here and slight poke there from time to time, always done in his courtly manner. Many thanks, Lynne.

Thanks go to my two technical reviewers, Asif Faiz and David Orr, who provided thoughtful, constructive, comprehensive comment on the draft manuscript. I confess I didn't always agree with them, and at times it was interesting reconciling their opinions—Asif being a former World Bank researcher and project manager, and David, a former county road engineer in the United States and now an LTAP director—but attempting to embrace their suggestions wherever possible has made the book that much stronger and more useful.

Comments by Gordon Keller on parts of the early manuscript are much appreciated.

To Taylor & Francis Group's Tony Moore go thanks for being so patient with me as I went through the distractions of the transition to an early retirement and moving house while writing the manuscript.

Finally, appreciation goes to my wife and best friend Chris, for putting up with me spending so many long hours lost to her absorbed in my study, and for periodically reminding me to "write for the ear" (as she reminds her Grade 7 students).

Robert A. Douglas

Author

Dr. Robert A. Douglas's passion for road engineering blossomed during his BASc studies in civil engineering studies (University of Windsor, Windsor, Ontario, Canada, 1976) and a PhD in geotechnical engineering (Southampton University, Southampton, United Kingdom, 1980). His professional life comprised both geotechnical engineering consulting work and an academic career. From 1979 to 1981, he worked for a geotechnical consulting firm in Southwestern Ontario and then, in 1982, moved to a Research Associateship at the Royal Military College, Kingston, doing early studies on geogrids in military roads. From 1983 to 2000, Robert held faculty positions in the Department of Forest Engineering and in the Department of Civil Engineering as the D.C. Campbell Chair for Highway Construction and Pavement Research, both at the University of New Brunswick (UNB), Fredericton, New Brunswick, Canada. The years spent in forest engineering at UNB focused on industrial forest haul roads.

In 2000, Dr. Douglas moved to the New Zealand National School of Forestry, University of Canterbury, Christchurch, where he was director of studies in forest engineering until 2006. This was followed by nine years' work in low-volume road engineering with a Canadian geotechnical engineering consulting firm back in Mississauga, Ontario, involving projects in such diverse locations as Northern Alberta, Ontario, and Quebec, the Canadian Arctic, and West Africa.

Dr. Douglas is now semi-retired, living in Windsor, Ontario, continuing low-volume road engineering work on a project-by-project basis.

About the book

As written by Michael Long, "Everything that sustains us—grown, mined, or drilled—begins [its journey to us] on a low-volume road (LVR)." And despite representing the largest proportion of any road network, roads of low-traffic volume have received short shrift in the engineering literature.

This book improves that situation. It covers the spectrum of the engineering of LVR from rural roads in developing countries, through farm-to-market roads in developed countries, to massive industrial haul roads carrying enormous trucks. Aspects of design, construction, and maintenance are presented. Topics covered include the following:

- Road classification, location, and geometric design
- Pavement concepts, materials, and thickness design
- Drainage, erosion and sediment control, and water crossings
- Slope stability
- Geosynthetics
- Road construction, maintenance, and maintenance management

Drawing on the author's experience of projects in rural Canada, the Canadian Arctic, Australasia, and West Africa, the book pulls together reference material from the United States, Canada, the United Kingdom, New Zealand, Australia, and South Africa. It presents examples of specifications and practices in those countries, using over 100 figures and 70 tables. It comfortably uses both metric and U.S. customary units throughout.

The book presents a complete treatment of the engineering of low-volume roads in a single resource.

Part I

Introduction

Chapter 1

Introduction to low-volume roads

> Low-volume roads are a lower order of worldwide land transport. They begin where animal track and walking trail end. They are the beginning of [the] world economy and are the life lines for rural communities. Everything that sustains us – grown, mined, or drilled – begins [its journey to us] on a low-volume road. Low-volume roads transcend language, culture, topography, and climate.
>
> Long [1.1]

1.1 SIGNIFICANCE OF LOW-VOLUME ROADS

Everything that sustains us—grown, mined, or drilled—begins [its journey to us] on a low-volume road (LVR). Could there be a more profound statement about as prosaic a thing as a road?

About 34 million kilometres of roads spread across the face of the earth, equivalent to a highway circling the equator more than 800 times. Only about 57% of the world's roads are sealed with bitumen, concrete, or stone [1.1]. The length of unsealed roads is extremely variable from region to region (Figure 1.1). Virtually all the world's unsealed roads, and about 85% of the sealed roads, carry a traffic volume of 1000 veh/day or less.

Beyond these publicly owned roads, there are uncounted millions of kilometres of private roads serving industries such as mining, agriculture, forestry, and the energy sector. The majority of these also have low traffic volumes.

Of the five dimensions of poverty,* the lack of accessibility is considered a key dimension [1.2]. Improving access helps to remove the constraints

* The other four are the lack of assets and income, physical debility, vulnerability to emergencies and disasters, and powerlessness [1.1].

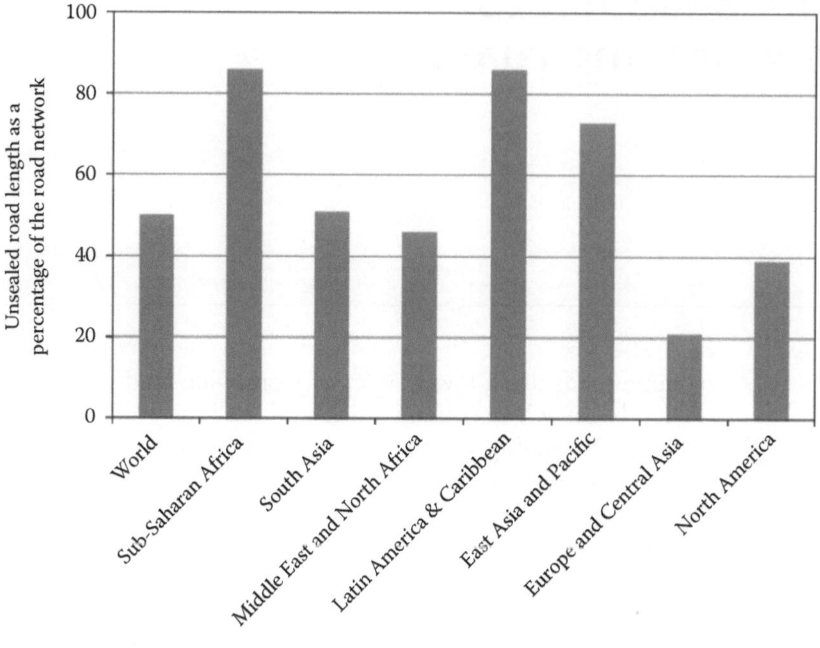

Figure 1.1 Proportion of unsealed road by global region—percentage of network. (From Faiz, A., The promise of rural roads—Review of the role of low-volume roads in rural connectivity, poverty reduction, crisis management, and livability, Transportation Research Circular E-C167, Transportation Research Board of the National Academies, Washington, DC, 2012.)

suffered by rural communities and offers the chance to improve livelihoods. Mapping the time needed to access a major city paints a fascinating picture (Figure 1.2). Improving access should be a major objective of rural development [1.1], and low-volume roads are the means for that development.

1.2 DEFINITION OF A LOW-VOLUME ROAD

But what exactly *is* a low-volume road (LVR)? How low is "low"? There isn't complete agreement. Faiz [1.1] sets the bar at an average daily traffic (ADT) of 1000 veh/day or less. The Transportation Association of Canada (TAC) defines an LVR as having an ADT of 200 veh/day or less [1.4], whereas the American Association of State Highway and Transportation Officials (AASHTO) geometric design guide for "very low-volume" roads [1.5] chooses an ADT of 400 veh/day or less.

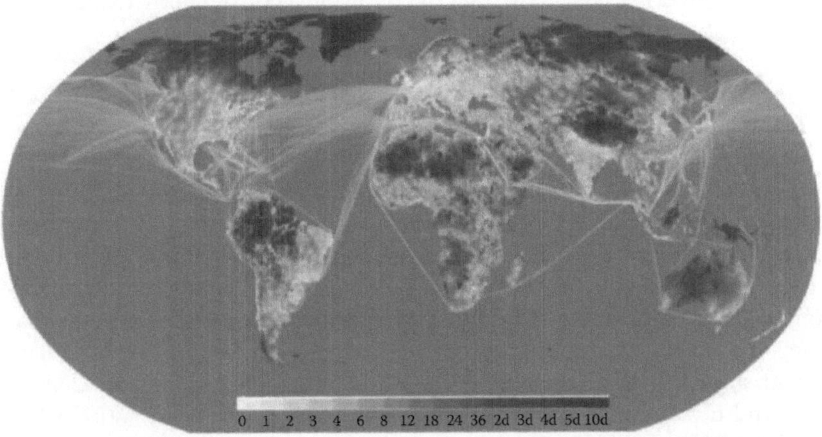

Figure 1.2 World accessibility, measured in hours or days travel time to a city of population 50,000 or more. (From European Commission, Travel time to major cities: A global map of accessibility, 2010, http://bioval.jrc.ec.europa.eu/products/gam/index.htm, visited February 2, 2015.)

There is rationale for accepting ADT 400 veh/day as the limit [1.6]:

- The peak hourly volume for rural roads of low volume is about 15% of the daily traffic.
- 15% of 400 is 60 vehicles.
- 60 veh/h translates to 1 veh/min.

Thus, "rush hour" on an LVR is equivalent to one vehicle passing by per minute. While the ceiling can range from an ADT of 200–1000 veh/day, depending upon topography, land use, traffic composition, population, and other socioeconomic factors, this book adopts 400 veh/day as the limit for LVR. As shown in Table 1.1, this definition captures approximately 70%–75% of the world's roads.

Table 1.1 Distribution of ADT on the global road network

Average daily traffic (ADT), veh/day	*% of global road network*
<400	70–75
<1000	80–85
<5000	98

Source: Faiz, A., The promise of rural roads—Review of the role of low-volume roads in rural connectivity, poverty reduction, crisis management, and livability, Transportation Research Circular E-C167, Transportation Research Board of the National Academies, Washington, DC, 2012.

LVR engineering has its own set of conditions, conveniently explained by the concept of "design environments."

1.3 DESIGN ENVIRONMENTS

The work presented in this book concerns two of the four road "design environments" (Figure 1.3) pertaining to traffic volume and design vehicle weight. Design environments so-defined provide a frame of reference for design, including the conditions, theoretical and empirical knowledge, standards, specifications, research and development for that type of road. They are set by the combination of the weight of the typical vehicle using a road, and the traffic volume on the road. Weights are simply divided into light and heavy, and volumes are divided into low and high. With a limit of 400 veh/day for LVR, the boundary between "low" volume and "high" volume is 400 veh/day. The boundary between "light" and "heavy" is arbitrary, and the concept really doesn't need to have specific values assigned anyway.

Note that the traditional LVR literature is concerned with only the lower left quadrant of the graph, whereas in most cases industrial haul roads (upper left quadrant) have low volume too. However, although the two design environments both have low volumes, they have vastly different design vehicle weights. A "low-volume road" such as a farm-to-market road in a developing country might have bicycles as the typical vehicle, whereas a mining road might be designed to support 600-tonne dump trucks. Clearly the pavement design aspects of the two design environments are vastly different.

The book covers both low-volume design environments.

1.4 APPROACHING THE SUBJECT OF LVR

LVRs truly receive international interest, and LVR experts are found all over the world. Therefore, wherever possible, the book presents and compares standards and practices prevalent in a number of countries, particularly the United States, Canada, the United Kingdom, Australia, New Zealand, and South Africa. Both SI metric and U.S. customary units are used, with the emphasis on the former. The international spelling of metre is adopted.

In the second main section after this Introduction, the book covers Road Design. The road design section generally follows the design process, covering, location, classification, geometric design and pavement concepts, materials, and design. Associated design topics, including drainage, erosion and sediment control, water crossings, slope stability, and design with geosynthetics, then follow. The third main section covers road construction and

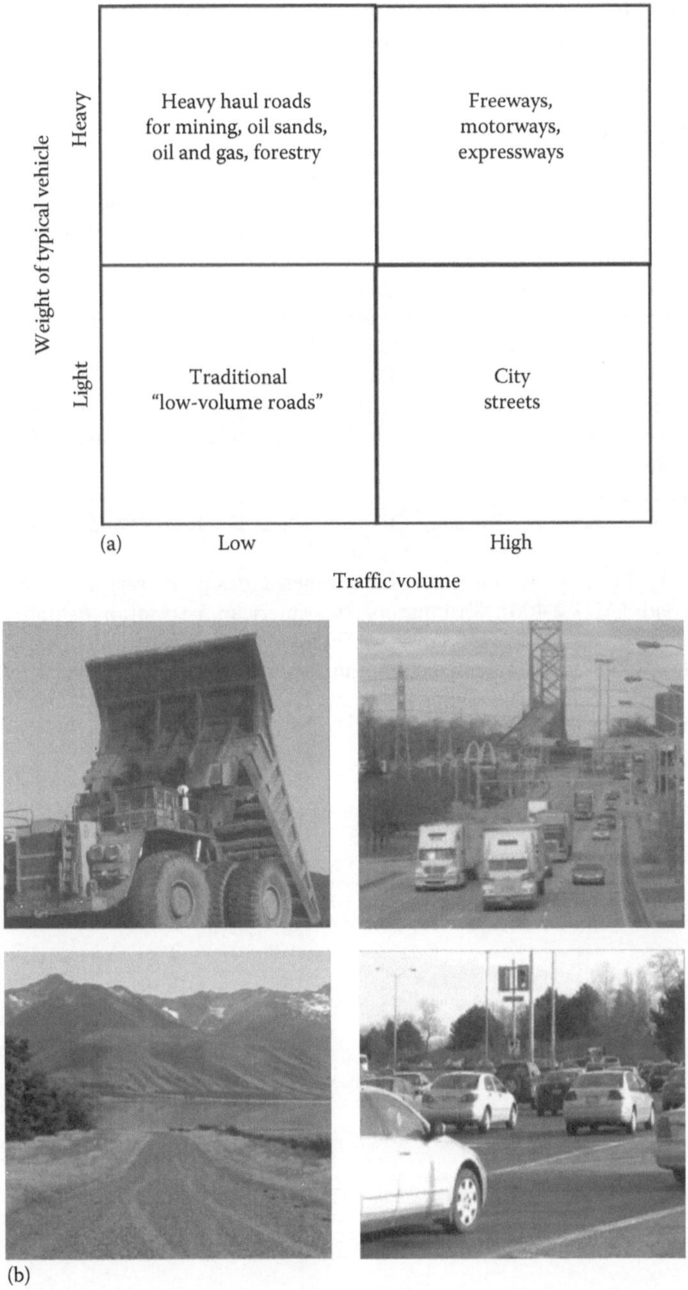

Figure 1.3 Road design environments: (a) roads typical of each design environment, (b) photographs of representative roads.

maintenance, with the final chapter examining maintenance management. Note that the aspects of environmental management are incorporated in various chapters where appropriate.

REFERENCES

1.1 Faiz, A. 2012. The promise of rural roads—Review of the role of low-volume roads in rural connectivity, poverty reduction, crisis management, and livability. Transportation Research Circular E-C167. Washington, DC: Transportation Research Board of the National Academies.

1.2 Chambers, R. 1980. Rural poverty unperceived: Problems and remedies. Background study for World Development Report 1980, Staff Working Paper No. 400. Washington, DC: The World Bank.

1.3 European Commission. 2010. Travel time to major cities: A global map of accessibility. http://bioval.jrc.ec.europa.eu/products/gam/index.htm, visited February 2, 2015.

1.4 RTAC. Manual of Geometric Design Standards for Canadian Roads, 1986 Metric Edition. Ottawa, Ontario, Canada: Roads and Transportation Association of Canada (RTAC, now TAC).

1.5 AASHTO. 2001. Guidelines for geometric design of very low-volume local roads (ADT < 400). Washington, DC: American Association of State Highway and Transportation Officials (AASHTO).

1.6 Irwin, L. 2013. Personal communication. Ithaca, NY: Cornell University, Local Technical Assistance Program.

Part II

Road design

Chapter 2

Road location

2.1 DEFINITIONS

Through road location, the designer chooses the best corridor for the road; through geometric design, the designer lays out the alignment of the route's centerline on the ground within the corridor. One first does the road location, selecting/defining a corridor, and then does the geometric design for the alignment within the corridor. This chapter covers road location, and then chapters on road classification and geometric design follow, classification being a prerequisite to geometric design.

2.2 A PHILOSOPHICAL EXERCISE

Road location is philosophical work. Writing well over a century ago in 1887, during the heyday of railroad construction, Wellington [2.1], quoted by Meyer [2.2], approached the metaphysical:

> ... there is nothing against which a locating engineer will find it necessary to be more constantly on his guard than the drawing of hasty and unfounded conclusions, especially of an unfavourable character, from apparent evidence wrongly interpreted. If his conclusions on reconnaissance are unduly favourable, there is no great harm done — nothing more at the worst will ensue than an unnecessary amount of surveying; but a hasty conclusion that some line is not feasible, or that further improvements in it cannot be made, or even sometimes — often very absurdly — that no other line of any kind exists than the one which has chanced to be discovered — these are errors which may have disastrous consequences.
>
> On this account, if for no other, the locating engineer should cultivate... what may be called an optimistic habit of mind. He should not allow himself to enter upon his work with the feeling that any country is seriously difficult, but rather that the problem before him is simply

> to find the line, which undoubtedly exists, and that he can only fail to do so from some blindness or oversight of his own, which it will be his business to guard against.
>
> For the reason that there is so much danger of radical error in the selection of lines to be surveyed (or rather, the lines not to be examined) it results that the worst errors of location generally originate in the reconnaissance. This truth, once grasped, the greatest of all dangers, over-confidence in one's own infallibility, is removed.

There is a strong message in this eloquent statement of the problem. Speaking down through the years, Wellington is cautioning against being too hasty, too willing to settle for the easiest road to survey. For the road locator pushing through thick vegetation and slogging through bog, all the while tormented by insects while surveying a new road, patience may run out fairly quickly, but it must be remembered that there could well be a much better final road "out there," which has been missed so far. There *is* a good road there; it just has to be found. It's senseless to rush the process. Big decisions are made early in road location, yet they come at a stage when the work being done itself is not very expensive, all things considered.

Road location is an optimization process. The alignment should have the least construction and maintenance costs possible, and it should minimize the disturbance to the natural environment. However, the alignment should result in the least vehicle operating costs possible too. These requirements oppose each other, hence the optimization exercise, and a balancing of the requirements to derive the minimum total cost system (road costs plus vehicle costs).

2.3 ROAD LOCATION CONSIDERATIONS

Numerous fundamental questions need to be answered by the road locator:

- What is the purpose of the road? Is it intended to provide direct access to the surrounding land or to provide through traffic?
- What is the nature of the terrain over which the road must pass? Is it flat, hilly, or mountainous? Does it vary along the proposed road? How much does it vary?
- Does the road pass through sensitive ecological or protected zones? Is there the potential for environmental harm?
- What road-building materials are available? Where are they located? Would quarrying or aggregate extraction cause undue damage to the natural environment?
- What construction expertise is available? For instance, is blasting experience available, and can it be called on economically? Or does the road alignment have to be set such that there will be no blasting?

- What road construction equipment is available? What can it do?
- What are the capabilities of the vehicles to use the road? What are their grade-climbing abilities, what are their economical operational maxima? How sharp a vertical curve can be accommodated without the vehicles becoming "hung up"? What are their minimum turning radii?

With so many competing requirements, and with so much riding on the outcome, road location is best done by people who have seen and thoroughly understand the whole road construction process, who can picture the final product, who understand how the final alignment will affect the performance of the vehicles to use the road, and who can fathom the economic, social, and environmental impacts of the road over its complete design life.

2.4 A PARADOX

A paradox occurs in road location. Early on in the process, the only cost to the road authority or owner is the wages of the locator or the survey crew, small costs compared to the cost of the overall project, and the subsequent road maintenance and vehicle operating costs. However, it is at that time that the most far-reaching, economically significant decisions are being made. The locator is deciding to run the road up this valley or that, over this ridge or the other one. The work is being done at the landscape scale.

Later in the process, the location is "locked in," and there is little that one can do to change things at that stage—the big decisions have already been made, and those remaining are relatively small. However, it is at that point that *costs* are large—the project already may be under construction, at which time the cost of making some alteration will be high; or the road may have already been constructed, with the authority, owner, and/or users stuck with whatever operating costs, high, or low, flow from the alignment that has been chosen.

Thus it makes sense for the locator to take time at the start, and not to rush the progress toward the final choice. A few extra hours or days cost little, but the "extra" time may result in great savings in the end. Suppose it takes an extra hour to find a new route that is 1 km shorter than previously contemplated: the 1 km of road saved might have cost thousands of times the extra hour of locator's time.

2.5 ROAD LOCATION PROCESS

The full, formal road location process consists of the following steps:

1. An office study
2. A field reconnaissance

3. A route survey
4. A location survey
    - Following this, in a contracted project, tender documents are prepared, the project goes out to tender, a contract is awarded, and construction begins
    - For an in-house project, design drawings and specifications are prepared, crews are mobilized, and construction begins
5. Construction surveys
6. As-built surveys

The work done in each of the steps is filled out more completely in the following paragraphs.

2.5.1 Office study

As the first step, the locator gathers and reviews as much relevant information as can be obtained regarding the proposed road, including remotely sensed data, maps, air photos, orthophotos, topographical sheets, soils reports, and geological maps. O'Flaherty [2.3] lays out the kinds of information that can be needed (Table 2.1) and also provides a comprehensive description of the mapping that is available in the United Kingdom.

As discussed earlier, the biggest decisions are made at this stage but they are made at the lowest cost. The end product of the office study is a decision to examine two or three candidate *corridors* each of perhaps 1 km (1/2 mile) width. Note that these are strips of land, rather than particular alignments.

2.5.2 Reconnaissance

Having selected two or three corridors, the next step is to go out and walk each one (or fly them in a helicopter or light plane, or use a drone), examining each to acquire a feeling for distances, gradients, ground conditions, obstacles, sources of aggregate, landing sites, water crossings, access, and so on. In temperate regions, this is best done in the autumn with the leaves off the deciduous vegetation when it is easiest to see the conditions, or early winter, before the snow is too deep for hiking. The end product of the reconnaissance is a decision to opt for one preferred corridor.

O'Flaherty [2.3] notes that in some cases this on-the-ground work needs to be carried out discretely to avoid land speculation and price blight.

2.5.3 Route survey

With a single corridor selected, the next step in the full road location process is to carry out an instrument survey of all the features within the corridor, where "instrument survey" means using whatever is appropriate in the spectrum running from Abney level, compass and chain, all the way up to total

Table 2.1 Information gathered in road location office study

General land survey
Site boundaries and building outlines
Topography and drainage
Permits and restrictions
Planning restrictions administered by regional and local authorities
Permits to construct over tunnels, mine workings
Mineral rights
Access
Temporary access for construction
Temporary and permanent permits to cross railways
Ground conditions
Geology
Presence of challenging soil conditions:
Permafrost, laterite, organic soils
Potential for flooding, landslides, erosion
Road-building material sources
Locations of pits and quarries
Quality, quantity
Sources of suitable waste materials from other processes
Drainage and sewerage
Names of authorities concerned
Locations, levels, and capacities of existing systems
Existing utilities
Names of authorities concerned
Locations and elevations of buried and overhead water, electrical, telecommunications, gas, oil pipeline, etc. services

Source: O'Flaherty, C.A., *Highways, the Location, Design, Construction, and Maintenance of Pavements*, 4th edn., Butterworth-Heinemann, Oxford, U.K., 553pp., 2002.

station survey equipment or GPS. It may be that the terrain information is provided by remotely sensed digital means such as LiDAR.* What results is a comprehensive, site-specific topographic map of the corridor itself.

A preliminary geotechnical investigation may be performed, to determine the soil conditions, groundwater conditions, depth to bedrock, and perhaps to evaluate nearby borrow sites. Lay [2.4] explains the benefits of identifying and mapping terrain units, which are "association[s] of terrain components forming... recognizable and distinct landform[s] and which all have characteristic physiography or soil and vegetation associations... ... A typical terrain unit might be 'ridge'." The road locator would avoid

* According to Wikipedia, LiDAR is either a portmanteau for "light" and "radar," or acronyms for "light detection and ranging," or "laser-illuminated detection and ranging." https://en.wikipedia.org/wiki/Lidar, visited April 7, 2015.

ridges for the road location because of the steep grades or possible shallow bedrock, but on the other hand, anticipate that ridge units could be the sites of future pits or quarries.

2.5.4 Location survey

Next, back in the office, the route survey is digested, and trial alignments are made, to arrive at the final alignment. Software is likely used at this stage, to facilitate the development of the alignment, and the determination of earthwork quantities.

The alignment is staked (pegged) out on the ground within the corridor, surveyed to confirm gradients, clearing and grubbing requirements, cut and fill quantities, and blasting quantities, if applicable. The preliminary geotechnical investigation may be filled out, to provide design information for pavements, culverts, and bridges.

In the contract situation, final designs for pavements, culverts and drainage features, and bridges are prepared. Contract documents including drawings and specifications are then drawn up, the project goes out to tender for the submission of bids, and a contract is awarded. For a road constructed in-house, design drawings and specifications are prepared, and construction crews are mobilized.

2.5.5 Construction surveys

After construction has begun, small surveys are carried out as construction proceeds. An example of their need is the requirement to determine cut and fill volumes for periodic payment certificates to be sent to the contractor.

2.5.6 As-built surveys

After the project has been completed, it is wise to do a complete field survey to document exactly how the road was finally constructed. There will have been changes, so the road won't exist exactly as shown on the construction drawings. To have a proper record to facilitate future, associated projects, there is a need to document the changes made in the current road project.

2.6 HOW MUCH OF THE FULL PROCESS IS ACTUALLY CARRIED OUT?

Of the full process, consisting of an office study, a field reconnaissance, a road survey, a location survey, construction surveys, and finally an as-built survey, how much is appropriate? One needs to justify the work from the economic viewpoint, and do only as much as is appropriate.

If the locator misses out steps or combines steps of the full, formal process, they should do so within a comprehensive knowledge of the full process. Such decisions should be made deliberately, rather than through ignorance of the full process. In the end, the work done in locating the road should match the scope, complexity, importance, cost, and expected life of the road.

2.7 STRATEGIES

A good road locator will have a number of strategies in mind when working toward the best alignment. Each has its advantages and disadvantages:

Stick to the valleys: In valleys, flat grades and reduced cut and fill requirements can be expected, but the road may be prone to floods and washouts, may have difficulty accommodating large radius horizontal curves, and may require a greater number of bridges and culverts.

Stay on the ridges: Generally, good drainage is found on ridge-top alignments, and often good construction materials are found there. However, ridge alignments often present steeper grades, and may make access to the surrounding land difficult.

Run across *the drainage pattern*: If the corridor crosses streams at right angles, shorter (less expensive) bridges will be needed, but a greater number of structures may be required, and their approaches may be awkward.

Run along the shoulders of hills (sidehill): With sidehill alignments, the locator can set gentle grades, but it may be difficult to achieve the required radius of horizontal curves, since those will be set by the topography.

Ignore the topography altogether: This is, of course, the simplest way to set the road alignment, but it may result in severe gradients, unsafe dips where stationary vehicles and other obstacles may be hidden, and poor vehicle performance.

So which strategy does the good locator adopt? The answer is, probably all of them, at one place or another along the corridor, taking advantage of each strategy's strengths, and minimizing its weaknesses.

2.8 WATER CROSSINGS

Water crossings represent major expenses in LVR construction and are the places where there can be the greatest potential for environmental damage. They are covered in detail in Chapter 10.

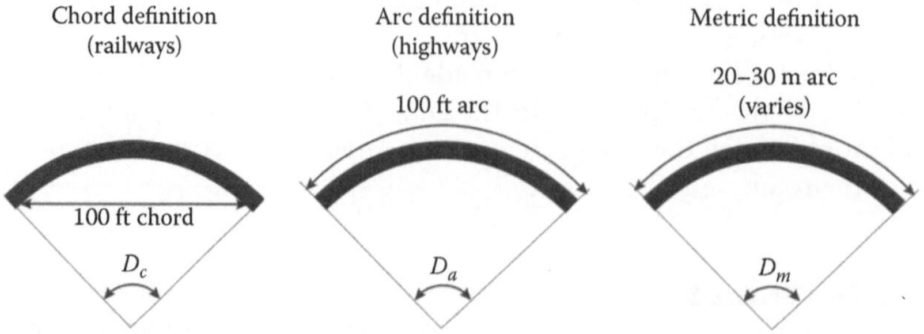

Figure 2.1 Plan view of chord and arc definitions of horizontal curvature (angles D_c, D_a, and D_m).

2.9 DEGREE OF HORIZONTAL CURVATURE

When working with older drawings of existing alignments, it is useful to note that there were three commonly used definitions of horizontal curvature. All three dealt with the degree of curvature, defined as the angle subtended by an arc or chord. However, the three definitions are different, as Figure 2.1 illustrates. More recent practice is just to identify the horizontal curve's radius in metres or feet.

2.10 RECOMMENDED GRADIENTS

A great deal of thought ought to go into the adoption of maximum values for the gradient of the roads being located. In addition, once policies are set as to what those values should be, they should be exceeded only in the most extreme cases, with a great deal of soul-searching (otherwise, why adopt them as "policy"?). Maximum gradients are presented in Table 4.6.

The effect of grades on haul vehicles must not be taken lightly. Figures 2.2a and b illustrate the impact a moderate grade (7%), which runs a theoretically infinite distance, would have on the speed of a haul truck with a mass to power ratio of 150 g/W (power/weight = 8 HP/ton). Such a ratio represents a truck with a gross mass of 45 tonnes (50 tons), and a full-throttle power of 300 kW (400 HP). As shown, with an entry speed of 50 km/h (30 mph), the truck would be slowed to just 18 km/h (11 mph) within 160 m (520 ft) of the start of the grade. Such a grade would represent a major bottleneck in a road network.

Section 4.12 covers the concept of critical length of grade in greater detail.

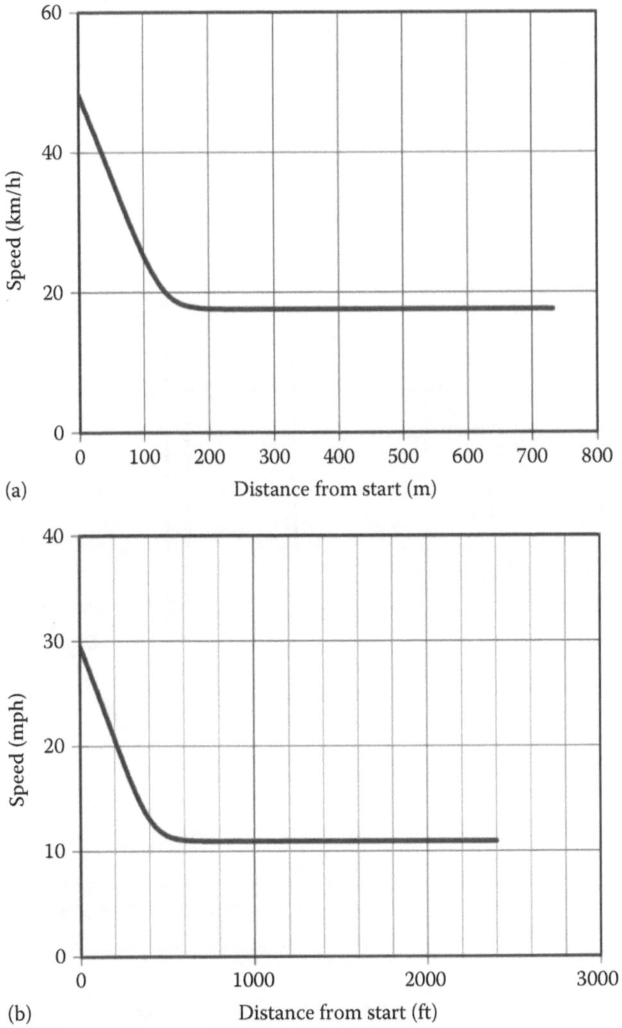

Figure 2.2 Theoretical speed profile for a truck with mass/power ratio 150 g/W (horsepower/weight ratio 8 HP/ton), climbing an infinite grade of 7%, (a) in metric units, (b) in U.S. customary units.

2.11 COMPUTER ASSISTANCE

The manual aspects of road location have been presented. However, digital data and computer software are becoming increasingly more sophisticated, and more useful.

Remotely sensed data such as air photos and satellite imagery can be invaluable in the office study and location survey stages. Digital elevation models (DEMs) derived from satellite data or LiDAR flights make the production of preliminary and final design drawings much more cost effective than on-the-ground surveys.

The spatial analysis techniques of geographic information systems (GIS) can be used to identify preferred routes (Figure 2.3). Using raster routines, a map of the area through which the proposed road will pass can be rated, grid cell by grid cell, for characteristics of relevance, such as

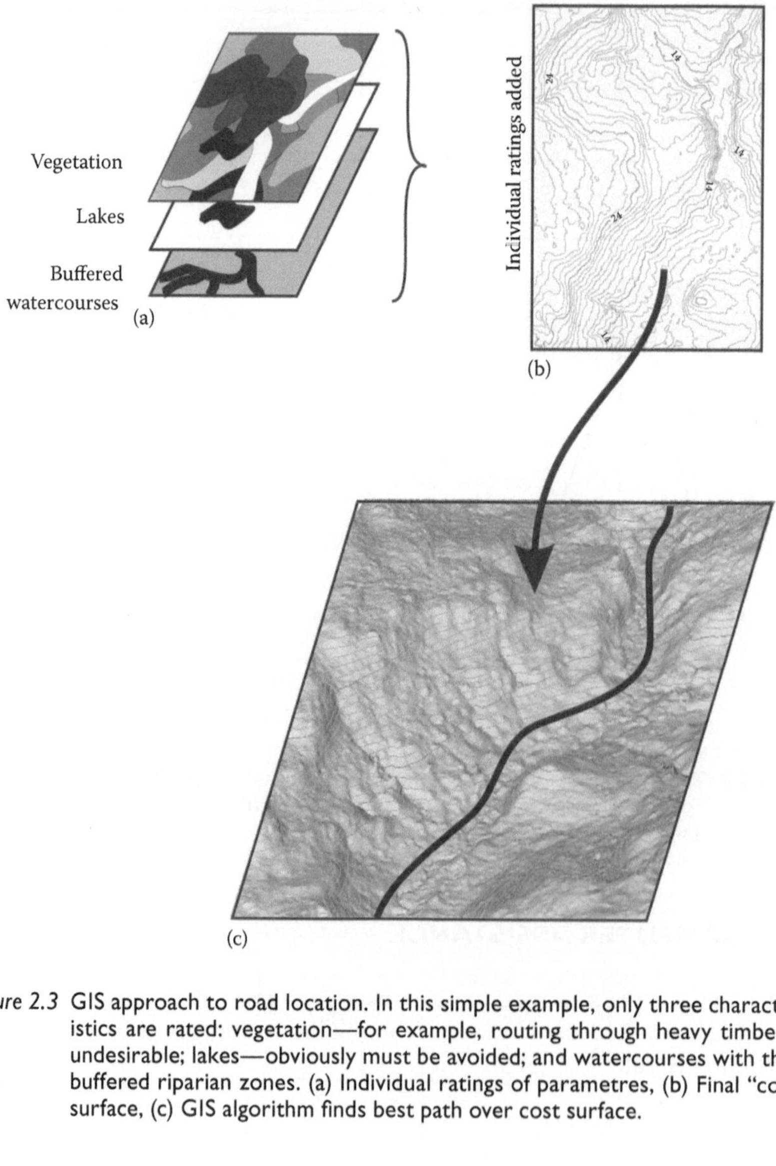

Figure 2.3 GIS approach to road location. In this simple example, only three characteristics are rated: vegetation—for example, routing through heavy timber is undesirable; lakes—obviously must be avoided; and watercourses with their buffered riparian zones. (a) Individual ratings of parametres, (b) Final "cost" surface, (c) GIS algorithm finds best path over cost surface.

- Land ownership
- Presence of water bodies
- Buffered watercourses
- Steep terrain
- Aspect of topography (slope direction)
- Boggy or soft terrain
- Areas of shallow bedrock
- Presence/absence of timber to be protected/harvested
- Lands of cultural and/or historical significance

For each characteristic (each theme, in GIS terminology), a "cost" is assigned (Figure 2.3a), where "cost" is a rating that reflects the undesirability of routing the road through that grid cell. For instance, routing the road through a lake will obviously incur a high "cost," so a very high "cost" would be assigned to all cells falling on lakes. The cost ratings for each theme might run from 0 (highly desirable) to 10 (road must avoid that grid cell). Weightings can be assigned to the costs for each theme to reflect its importance.

The weighted costs are summed for each grid cell (Figure 2.3b), to produce the total cost of routing the road through each grid cell. The total costs are then portrayed as a surface (Figure 2.3c), where the "valleys" in the surface—the areas of minimal cost—represent the most desirable places to route the road. Three-dimensional routines in the GIS system can be used to find the minimum cost path over the cost surface, between two end points.

REFERENCES

2.1 Wellington, A.M. 1915. *Economic Theory of the Location of Railways*. Toronto, Ontario, Canada: John Wiley & Sons. Inc.

2.2 Meyer, C.F. 1967. *Route Surveying*. Scranton, PA: International Textbook Co., pp. 6, 7.

2.3 O'Flaherty, C.A. 2002. *Highways, the Location, Design, Construction, and Maintenance of Pavements*, 4th edn. Oxford, U.K.: Butterworth-Heinemann, 553pp.

2.4 Lay, M.G. 2009. *Handbook of Road Technology*, 4th edn. New York: Spon Press, 933pp.

Chapter 3

Road classification

3.1 INTRODUCTION

The entry point to road geometric design is the classification of the road. Before covering geometric design, road classification will be discussed.

There are probably thousands of classification systems. Biologists deal with the taxonomy of animals and plants. Geologists classify rocks. Geotechnical engineers classify soils. Soil scientists do too, but with different systems. Sports organizations sort teams into leagues. And so on.

There are a number of common points to be made about classification systems:

- In every case, there is a sorting of the population into groups of individuals with common characteristics.
- There is some description of those characteristics, either explicit or implied.
- There is the intention to convey a common picture of the characteristics of an individual in one of the groups to everyone familiar with the classification system.
- Predictions of the behavior of an individual can be made based on the classification of that item.

The last two points are the most important. The reasons to classify anything are to have a short-hand description, which provides a clear picture of what that thing is like, and to be able to predict how it will behave. The same is true for the classification of roads.

Functional classification systems exist for roads in developed countries, with classes such as interstate, primary, collector, and local. While LVR usually fall into the lowest one or two classes in such systems in developed countries, because of the preponderance of two- and three-wheeled vehicles, bicycles, nonmotorized traffic and pedestrian traffic, even primary roads may be classified as LVR in terms of an ADT measure that only accounts for motor vehicles.

The following sections outline the road classification systems for LVR prevalent in the United States, Canada, the United Kingdom, South Africa, and Australia.

3.2 THE U.S. SYSTEM

The U.S. system is provided in the *Guidelines for Geometric Design of Very Low-Volume Local Roads* [3.1] known as the "Little Green Book." Rural and urban roads with AADT less than 400 veh/day are classified according to function as shown in Table 3.1.

Table 3.1 AASHTO road classifications

Classification	*Description*
Rural roads	
Major access roads	Serve a dual function: access to abutting land and connecting service to other roads
Minor access roads	Almost exclusively used for access to abutting land—roads are typically short—may be dead ends—speeds are low
Industrial/commercial access roads	Provide access to factories or commercial outlets—high proportion of truck traffic—roads are typically short and dead ended
Agricultural access roads	Provide access to fields—traffic often consists of large, heavy, and or slow agricultural equipment
Recreational and scenic roads	Serve specialist land uses such as parks and campsites—roads have low truck traffic but typically do cater to large, slow recreational vehicles
Resource recovery roads	Serve mining or logging operations—traffic has a very high proportion of large, heavy resource industry trucks
Urban roads	
Major access streets	Serve a dual function: access to abutting land and connecting service to other roads—typically shorter than rural equivalents
Urban residential streets	Access to single and multiple family residences—large trucks are rare
Industrial/commercial streets	Serve developments that generate a substantial number of large and/or heavy trucks—typical function is to provide connection between a factory and highway

Source: AASHTO, *Guidelines for Geometric Design of Very Low-Volume Local Roads (AADT < 400)*, American Association of State Highway and Transportation Officials (AASHTO), Washington, DC, 2001.

3.3 THE CANADIAN SYSTEM

As shown in Table 3.2, in the Canadian system [3.2], roads are simply designated as "LVR" with the design speed appended. Eight designations are permitted. The table indicates the recommended limits for gradients, sight distances, superelevation (banking), and curve radii.

3.4 THE UK SYSTEM

Overseas Roadnote 8 [3.3] presents a system of classification for all roads. The access categories pertain to volumes less than 400 veh/day. Table 3.3 shows the road classifications, traffic volumes, surface type, dimensions, gradients, and design speeds.

3.5 THE SOUTH AFRICAN SYSTEM

An excellent compilation of road classifications is provided in the South African Road Classification and Access Management Manual [3.4]. The manual notes the prevalence of six-class systems. The publication does not identify low-volume roads specifically, however, these would generally fall into Rural Classes 4 and 5, Collectors and Local Roads, respectively, when classed according to function (Table 3.4).

3.6 THE AUSTRALIAN SYSTEM

Giummarra [3.5] examined road classification systems across Australia and abroad, and found that most rural LVR fall into Class 5 of the systems in place in Australia. For LVR, he recommended subdivision of Class 5 into five more classes, as shown in Table 3.5.

Table 3.2 TAC low-volume road classifications

Classification	*Design speed (km/h)*	*Maximum gradient (%)*	*Minimum stopping sight distance[a] (m)*	*Maximum superelevation (m/m)*	*Minimum radius of curve (m)*
LVR30	30	11–16	30	0.08	30
LVR40	40	11–15	45	0.08	50
LVR50	50	10–14	65	0.08	80
LVR60	60	10–13	85	0.08	120
LVR70	70	9–12	110	0.08	170
LVR80	80	8–10	140	0.08	230
LVR90	90	7–9	170	0.08	300
LVR100	100	6–8	200	0.08	390

Source: RTAC, *Manual of Geometric Design Standards for Canadian Roads*, Roads and Transportation Association of Canada (RTAC, now TAC), Ottawa, Ontario, Canada, 1986.

[a] Applicable to two-lane roads and one-lane one-way roads.

Table 3.3 Overseas Road Note 6 road classifications

Road function	*Design class*	*Traffic volume (veh/day)*	*Surface type*	*Width*		*Maximum gradient (%)*	*Design speed by terrain*		
				Running surface (m)	*Shoulder (m)*		*Mount (km/h)*	*Rolling (km/h)*	*Level (km/h)*
Access	D	100–400	Paved/unpaved	5.0	1.0[a]	10	50	60	70
	E	20–100	Paved/unpaved	3.0	1.5[a]	15	40	50	60
	F	<20	Paved/unpaved	2.5–3.0	Passing places	15–20	n/a	n/a	n/a

Source: TRL, *Overseas Road Note 6*, Overseas Unit, Transport and Road Research Laboratory (TRRL, now TRL), Crowthorne, U.K., 1988.

[a] For unpaved roads where the running surface is gravelled, the shoulder would not normally be gravelled, however, for Class D roads, consideration should be given to gravelling shoulders if shoulder damage occurs.

Table 3.4 South African road classifications (COTO [3.4])

Rural road class	*FHWA description*	*Percentage of veh-km*	*Percentage of road length*
R1, R2	Principal arterials	30–55	2–4
R1, R2, R3	Principal plus minor arterials	45–75	6–12
R4	Collectors	20–35	20–25
R5	Local roads	5–20	65–75

Table 3.5 Proposed Australian road classifications

Road class	*Class type*	*AADT (veh/day)*	*Design speed (km/h)*	*Running surface width (m)*	*Description*
5A	Primary road	>100	50–80	7	Main traffic movement through a region
5B	Secondary road	50–100	30–70	5.5	Collects and distributes traffic from local areas
5C	Minor road	20–50	20–60	4	Links areas which are traffic generators to secondary or primary roads
5D	Access track or road	<20	<20–40	4	Provides access to low-use sites
5E	Rough track	<10	n/a	3	Primarily for four-wheel drive vehicles

Source: Giummarra, G., Road classifications, geometric designs and maintenance standards for low-volume roads. Research Report ARR354. ARRB Transport Research Ltd., Vermont South, Victoria, Australia, 2001. Used with permission of ARRB group Ltd.

3.7 A NOVEL SYSTEM

All of the systems described earlier are closed systems; there is a finite set of classifications into which any road must go. Sometimes the fit is not good.

In an open system devised by Paterson et al. [3.6], the important engineering characteristics of the road are directly embedded in its classification name. The road designations have the following fields:

Al axle loading: expressed as the maximum single axle load expected to occur over the life of the road (has an impact on the design of the road structure)

S speed: expressed as the *minimum* speed for the design vehicle which can be tolerated (a reflection of the maximum gradients allowed)

A availability: Y = year round

A = all weather (except during season of thaw and subsequent recovery)
W = winter only (i.e., only when frozen)
D = dry weather only
T traffic volume: number of equivalent single axle loads (ESAL) per day
L design life: the number of years until major rehabilitation or abandonment of the road

The system is called the AL-SAT-L system after its field names. In the system, a road classed for example as 100 kN—15 km/h—D—10 ESAL—1 would be one designed to carry a maximum axle loading of 100 kN/axle, with gradients flat enough that a speed of at least 15 km/h could be maintained. The road would be available for use during dry weather only, is designed to carry 10 ESAL/day, and would have an expected life of 1 year.

A completely open classification system results from taking the Al-SAT-L approach. Any road can be classed, and its relevant engineering characteristics can be conveyed by the classification designation. The system was designed specifically for logging roads, and although it could also be used for mining, oil sands, and oil and gas, it has never been implemented routinely.

REFERENCES

3.1 AASHTO. 2001. *Guidelines for Geometric Design of Very Low-Volume Local Roads (AADT < 400).* Washington, DC: American Association of State Highway and Transportation Officials (AASHTO).
3.2 RTAC. 1986. *Manual of Geometric Design Standards for Canadian Roads.* Ottawa, Ontario, Canada: Roads and Transportation Association of Canada (RTAC, now TAC).
3.3 TRL. 1988. *Overseas Road Note 6.* Crowthorne, U.K.: Overseas Unit, Transport and Road Research Laboratory (TRRL, now TRL).
3.4 COTO. 2012. *South African Road Classification and Access Management Manual.* Pretoria, South Africa: Roads Coordinating Body of the Committee of Transport Officials (COTO), The South African National Roads Agency Ltd.
3.5 Giummarra, G. 2001. Road classifications, geometric designs and maintenance standards for low-volume roads. Research Report ARR354. Vermont South, Victoria, Australia: ARRB Transport Research Ltd.
3.6 Paterson, W.G., McFarlane, H.W., and Dohaney, W.J. 1969. A proposed forest roads classification system. Woodlands Reports W.R. No. 20. Pointe Claire, Quebec, Canada: Pulp and Paper Research Institute of Canada, 47pp.

Chapter 4

Geometric design

4.1 INTRODUCTION

Geometric design deals with the details of the alignment of roads, such as the gradients and horizontal curve radii. It is carried out to ensure the safe, smooth flow of traffic. Consideration is also paid to the fast, efficient, economical movement of vehicles, although for public projects, accurately quantifying the savings due to good geometric design gets a little tricky.

This chapter covers the geometric design of LVR in developed countries, where geometric design satisfies the needs of motorized traffic. In developing countries, LVR may have low levels of motorized traffic, and the traffic stream has high proportions of pedestrians, bicycles, carts, other nonmotorized traffic, agricultural tractors with trailers, motorcycles, and motorized three-wheeled vehicles. Advice on the appropriate design of roads and tracks for these conditions can be found in the developing country literature (see for example [4.1,4.2]).

There are a number of sources of design information including the following:

- United States [4.3,4.4]
- Canada [4.5]
- United Kingdom [4.6]
- Australia and New Zealand [4.7–4.9]
- South Africa [4.10]

The authors of these references are careful to point out that their publications are only intended to serve as guides:

> "The design engineer's fundamental responsibility is to exercise professional engineering judgment on technical matters in the best interests of the public including safety and cost effectiveness. Standards are available to assist in making those judgments, but should not be used as a substitute...

> ...Because of the essential need to apply design standards with judicious care and proper consideration for prevailing circumstances, the Manual is not suitable for adoption as a design code in a regulatory sense nor can it be regarded as a set of minimum specifications to protect public safety." [4.10, "Guidance for Users"]

The final responsibility for the design lies with the designer. The same is true for the material presented in this chapter, particularly since a great deal of what is presented must be considered as preliminary for LVR design, until research on the different conditions posed by LVR can catch up.

Geometric design standards concern all aspects of the geometric design of roads, from curves to intersections, even to things like lighting and noise abatement. However, this chapter will be restricted to the main elements of geometric design relevant to LVR:

- Sight distances
- Horizontal curves
- Side clearances around curves
- Gradients
- Vertical curves
- Roadway cross sectional details

The details are usually presented separately as horizontal, vertical, and cross sectional alignment elements, but it must be remembered that in the design, the elements must be considered simultaneously.

When considered as a rigid body, a vehicle has the six degrees of freedom defined in Figure 4.1. Articulation at trailer connections adds more. The motions the vehicle goes through are controlled partly by the geometric design of the road alignment, which determines the path through space that the vehicle follows (assuming the vehicle is intended to stay on the road!). Therefore, a smoothly varying path through space will make the trip safer, faster, and more economical. The comparison of the line followed by an airplane, as it flies through the air (climbing, banking, turning), to that of a car or truck as it follows the alignment of a road, is a good one.

4.2 DESIGN VEHICLES AND THE GEOMETRIC DESIGN OF HAUL ROADS

Geometric design is based on a design vehicle. This is a fictitious vehicle characterized by a number of assumed parameters, such as the power to weight ratio, or the distance from the pavement surface to the driver's eye. For most of the geometric design of conventional, public roads, the design

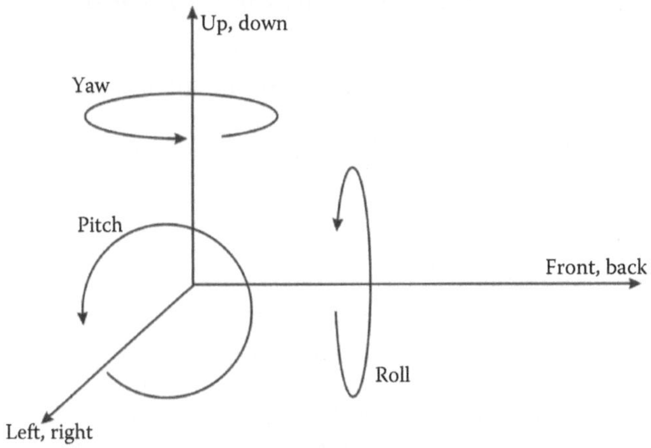

Figure 4.1 The six degrees of freedom of a vehicle, when considered as a rigid body.

vehicle is a passenger car. The designer develops an alignment that satisfies the requirements of a passenger car.

Haul roads stand in contrast to public roads in a number of ways. Designers of public roads must deal with the serious challenges of estimating what the composition of the traffic stream—the proportions of cars, vans, buses, and heavy trucks—will be, and how the traffic volume—the number of vehicles per day using the road—will grow, over the life of the road. Fortunately, however, they have the support of a large body of information, including design standards and methods, data and case histories, to work with.

Haul road designers, on the other hand, usually know very accurately what the traffic volume will be, and how it will vary over time. Knowing exactly which vehicles will be used and what their capacities are, they need only consult the operating plans for the facilities to be served by the haul roads to determine traffic volumes. The composition of the traffic stream need not be guessed.

However, they must work with *two* design vehicles, not just one.

> *Two* important types of vehicles use haul roads: (1) the passenger cars and/or pickup trucks (utes) of supervisory and technical staff, and (2) the heavy trucks doing the hauling.

If it is recognized that the haul road design environment is different, and further, that one of its characteristics is a bimodal traffic composition, a clear picture can be assembled, indicating which conventional geometric design standards can be borrowed and used directly, which standards must be modified and finally, what gaps must be filled in with further research.

4.2.1 Geometric design controls for haul roads

Table 4.1 sets out the controlling factors for those aspects of the geometric design of haul roads to be considered here. In some cases, the small, fast vehicles will control the element of geometric design under consideration. In other cases, the large, slow vehicles will control. In still other cases, the design element is not based on either design vehicle rather it depends on some other completely different aspect.

Stopping sight distance: An issue for haul roads, and more generally for gravel-surfaced roads, is the modeling of the braking of very heavy haul trucks running on unbound surfaces (roads with untreated soil surfaces, graveled surfaces, or surfaces of crushed rock). The conventional model of sliding friction between the road surface and the vehicle's tires, which are assumed to be locked and sliding along the road surface during braking, is not valid for the heavily loaded haul truck running on an unbound surface. The granules of the road surface material can separate and roll under the tires during braking or lateral sliding of heavy trucks. Under severe, "panic" braking, the weight of the heavy haul truck is enough to prevent wheel lockup and sliding. The friction in the brakes themselves, and brake component efficiency become determining factors, invalidating the assumed tire-sliding-on-road model. Also, small piles of loose aggregate can be plowed up ahead of the tires if the wheels do indeed lock up, complicating the modeling even more.

With the rolling-granules and/or unlocked-braked-wheel phenomena occurring, the braking distances can be greater than those calculated using the sliding friction approach adopted in conventional geometric design. As well, the characteristics of lateral sliding while negotiating horizontal curves could be quite different from what is expected.

The stopping sight distance—the distance between an object or hazard on the road for which the vehicle must be stopped, and the point at which it comes into the view of the driver—may thus be controlled by the fast light vehicles, where speeds are high and the driver sits relatively close to the road surface. On the other hand, it may be controlled by the slow, heavy haul trucks, where despite the higher vantage point afforded the driver and the slower design speed, the comparatively poorer braking characteristics of the much heavier vehicle result in a longer braking distance.

Horizontal curves: For the design of circular horizontal curves, superelevation—the maximum banking of the road cross section—is governed by the behavior of vehicles during icy or slippery conditions. The radius of the curves, however, may be governed by the need for safety of the fast, light vehicles using the road, which must not be allowed to slide off laterally, or by the possibility of the heavily loaded

Table 4.1 Controls on the geometric design of haul roads

Element	*Control on design element, comments*			*Remarks*
	Pickup truck (ute)	*Haul truck*	*Other*	
Stopping sight distance	√ High speed may control	√ May control		Research needed on braking distance for heavy haul trucks, braking with wheels continuing to rotate
Horizontal alignment				
Lateral friction				Assumed independent of vehicle: behavior of unbound surfaces?
Maximum superelevation			√	Icing of road surface, slippery surface when wet, comfort of operators
Circular curves	√ High speed may control	√ Roll-over may control		
Lateral clearance	√ Stopping sight distance	√ Stopping sight distance?		
Vertical alignment				
Minimum gradient			√	Drainage
Maximum gradient		√		
Critical length of grade		√		
Crest vertical curves	√			
Sag vertical curves	√	√		Same for both vehicles
Cross-sectional elements				
Cross section		√ Lane width	√	Drainage

haul trucks, with much higher centers of gravity, rolling over while attempting to negotiate the curve.

Maximum gradients: Maximum gradients and critical lengths of grade are clearly governed by the climbing ability of the slow, heavy haul trucks, with their comparatively high mass/power ratios.

Crest vertical curves: The design of crest vertical curves, where the road goes over a hill, is governed by the required stopping sight distance of the small, light, fast vehicles, with the driver sitting comparatively close to the road surface.

Cross sectional elements: Roadway cross sectional elements, particularly the width of the running surface, are obviously governed by the width of the large haul trucks.

In the next sections, the derivation of design guidelines for each of the geometric design elements listed in Table 4.1 will be covered in detail for both types of design vehicles, and the practices in a number of countries outlined.

4.3 STOPPING SIGHT DISTANCE

For every point along the road alignment, the designer must provide enough sight distance for the driver to see an object or hazard on the road and to be able to stop for it in time. Adequate stopping sight distance must be provided continuously, every metre along the road.

If the driver comes upon such a situation (a fallen log on the roadway, a culvert or a bridge washed out), they must first see the hazard, recognize it as a hazard, determine that the vehicle will have to be stopped, and initiate the action of stepping on the brake pedal. A time interval passes, while the driver decides that stopping is necessary, during which the vehicle continues to travel along at unchanged speed. During this time, the *brake reaction distance* is covered. Once the brakes have been applied, the vehicle is brought to a stop over the *braking distance*. These two distances, the brake reaction distance plus the braking distance, added together give the minimum stopping sight distance.

The brake reaction distance is affected by a number of factors, including the physiological condition of the driver (age, reaction time, sobriety, reactions to drugs, fatigue, distractions, fitness) and the site conditions (ambient light, dust, rain, snow, fog). A standard time of 2.5 s has been widely adopted for the brake reaction time, being about the 95th percentile of the distribution of brake reaction times [4.3].

Two approaches to calculating braking distance are seen in the literature. Either a coefficient of longitudinal sliding friction (f in the following equations) is assumed and the braking distance calculated for the vehicle sliding as a rigid body to a stop, or a coefficient of deceleration is assumed, and the distance covered while decelerating to a stop calculated. The deceleration rate is $f \times g$ in the equations following, where g is the acceleration due to gravity, 9.81 m/s^2, or 32.2 ft/s^2. The latter approach is more elegant, as it permits the consideration of antilock braking systems, heavy trucks braking with unlocked wheels, and poorer braking systems.

Here are the equations:

(a) Brake reaction distance:

$$d_{br} = V\,t \tag{4.1}$$

where

d_{br} is the brake reaction distance
V is the initial speed
t is the brake reaction time, usually taken as 2.5 s.
Note: the units must be consistent (m, m/s, and s; or ft, ft/s, and s)

For mixed metric units:

$$d_{br} = 0.278 V t \tag{4.2}$$

where

d_{br} is the brake reaction distance (m)
V is the initial speed (km/h)
t is the brake reaction time (s)

For mixed U.S. customary units:

$$d_{br} = 1.47\ V\,t \tag{4.3}$$

where

d_{br} is the brake reaction distance (ft)
V is the initial speed (mph)
t is the brake reaction time (s)

(b) Braking distance, consistent units:

$$d_b = \frac{V^2}{2fg} \tag{4.4}$$

or

$$d_b = \frac{V^2}{2a} \tag{4.5}$$

where

d_b is the braking distance
V is the initial speed
f is the coefficient of friction between tires and road surface
g is the acceleration due to gravity
a is the assumed rate of deceleration during braking
Note: the units must be consistent (m, m/s, and s; or ft, ft/s, and s)

For mixed metric units:

$$d_b = \frac{V^2}{254f} \tag{4.6}$$

where

d_b is the braking distance (m)
V is the initial speed (km/h)
f is the coefficient of friction (unitless)

For mixed U.S. customary units:

$$d_b = \frac{1.075V^2}{a} \tag{4.7}$$

where

d_b is the braking distance (ft)
V is the initial speed (mph)
a is the coefficient of deceleration (ft/s^2)

4.3.1 Stopping sight distance in practice

Table 4.2 presents a summary of stopping sight distances by country. The typical practice in the United States [4.3] is to use the deceleration approach and Equations 4.3 and 4.7. Typical practice in Canada [4.5], Australia and New Zealand [4.9], and South Africa [4.10] is to use the friction approach (though longitudinal friction is stated as deceleration rates in the Australian and New Zealand reference) and Equations 4.2 and 4.6.

4.4 COEFFICIENT OF SIDE FRICTION

As a vehicle negotiates a horizontal curve, the superelevation (banking) of the curve and the sideways friction that the tires can develop against the road surface must be enough to keep the vehicle on the road at the current speed, as shown schematically in Figure 4.2.

Table 4.2 Summary of stopping sight distances by country (for sealed roads unless otherwise noted)

Speed (km/h)	*Speed (mph)*	*U.S.*		*Canada*		*UK*		*Australia and New Zealand*		*Australia and New Zealand*		*South Africa*	
		(m)	*(ft)*	*(m)*	*(ft)*	*(m)*	*(ft)*	*(m)*	*(ft)*	*(m)*	*(ft)*	*(m)*	*(ft)*
30	20	30	95	30	100	25	80						
50	30	55	165	65	210	50	160	45	150	50	160	65	210
70	45	90	300	110	360	85	280	85	280	100	330	95	310
80	50	110	350	140	460			115	380	135	440	115	380
100	60	155	470	200	660	160	520	170	560	195	640	155	510
	Remarks	For 250–400 veh/day								For unsealed roads			
	References	[4.3]		[4.5]		[4.6]		[4.9]		[4.9]		[4.10]	

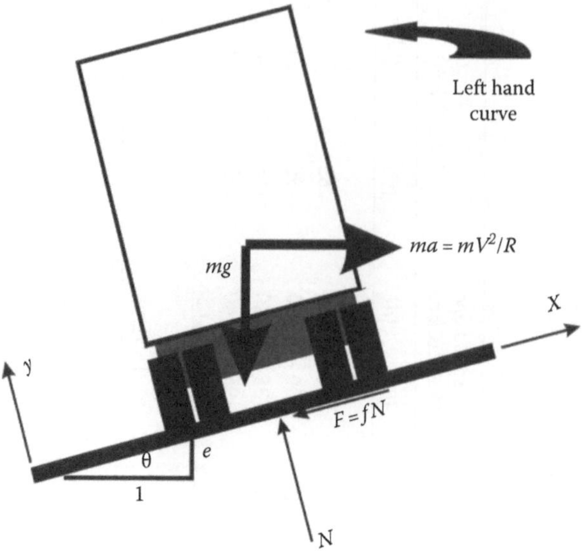

Figure 4.2 The equilibrium of a truck assumed not to roll over, on a superelevated curve.

Design coefficients of side friction less than those used for stopping sight distance calculations are adopted, because the consequences of "failure" (an overestimation by the designer of the friction actually available to the vehicle) while negotiating a horizontal curve are far greater than those for "failure" while braking. If the actual friction is less than the value the designer has used in the design of the horizontal curve, the vehicle will slide sideways off the road surface, inexorably crashing into whatever is beyond the running surface, at a speed close to the design speed of the roadway. Failure could have disastrous consequences. On the other hand, for straight ahead stopping, the actual friction being less than the value the designer has used in determining the required stopping sight distance in a panic stop will most likely cause less dire consequences than sliding sideways off the road in a curve will. The vehicle will only strike the object or hazard some time after first seen, after the driver presumably will have been able to reduce speed to some extent. It may also be possible to steer around the hazard. The consequences of "failure" in that context would be comparably less.

Thus, the values adopted for the coefficient of side friction are reduced from those used in stopping sight distance calculations, injecting a higher inherent factor of safety into the calculations for horizontal curves. Values used in a number of countries are shown in Table 4.3.

Table 4.3 Summary of design side coefficients of friction by country (for sealed roads unless otherwise noted)

Speed (km/h)	*Speed (mph)*	*U.S.*	*U.S.*	*Canada*	*UK*	*Australia and New Zealand*	*South Africa*
30	*20*	0.17	0.31	0.17	0.33		0.33
50	*30*	0.16	0.21	0.16	0.25	0.12	0.25
70	*45*	0.14	0.16	0.15	0.20	0.10	0.20
80	*50*	0.14		0.14		0.10	
100	*60*	0.12		0.13	0.15	0.09	0.15
Remarks		Higher volume roadways	Low speed urban streets			For unsealed roads	
References		[4.3]	[4.3]	[4.5]	[4.6]	[4.9]	[4.10]

4.5 MAXIMUM SUPERELEVATION

Superelevation is the banking used in horizontal curves. It is usually given as a dimensionless ratio or percentage of the road surface's rise/run, taken at right angles to the road axis. This alignment element is independent of vehicle type and is controlled by the comfort of the driver or the climatic or weather conditions to which roads are subjected. The general consensus [4.6] is that the maximum superelevation for driver comfort is 10%. Beyond that, the sideways force becomes uncomfortable. The maximum value of superelevation also must be set low enough that during icy, wet, or slippery conditions, a vehicle is not presented with so much banking that it slides sideways toward the center of the curve while either negotiating it at less than design speed, or when attempting to start up from a dead stop on the curve. Considering icy conditions, conventional Canadian practice sets the maximum value of superelevation at 8% [4.5].

Note that superelevation can interfere with the drainage of the road surface, especially in hilly or mountainous terrain. See Section 8.2 for more discussion.

4.6 HORIZONTAL CURVES

Having discussed side friction factors and the maximum superelevation, it is now possible to consider the design of horizontal curves. These usually consist of an arc of a circle, where the radius is constant. To provide a smooth transition from the straight ahead tangent alignment with the normal cross section to the superelevated circular curve, a spiral transition curve, where the radius and superelevation vary smoothly as the vehicle goes along, may be placed between the tangent alignment and the circular curve (Figure 4.3). The mirror image (circular curve to spiral to tangent) is provided at the exit of the horizontal curve.

The equilibrium equations for the vehicle on the curve shown in Figure 4.2 can be used to develop an equation for the minimum radius of the circular portion of the curve.

For mixed metric units:

$$R_{min} = \frac{V^2}{127\ (e_{max} + f)} \qquad (4.8)$$

where

R_{min} is the minimum radius of the curve (m)
V is the vehicle speed (km/h)
e_{max} is the superelevation in the circular curve (m/m)
f is the coefficient of side friction (unitless)

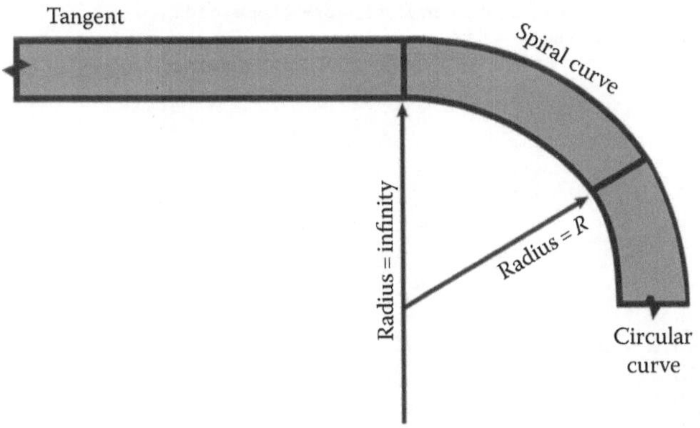

Figure 4.3 Plan view of a horizontal curve with spiral transition.

For mixed U.S. customary units:

$$R_{min} = \frac{V^2}{15\,(e_{max} + f)} \tag{4.9}$$

where

R_{min} is the minimum radius of the curve (ft)
V is the vehicle speed (mile/h)
e_{max} is the superelevation in the circular curve (ft/ft)
f is the coefficient of side friction (unitless)

Design tables can be arranged, as in Figure 4.4 [4.5].

4.6.1 Trucks with high centers of gravity: Rollover

As it had been noted in Table 4.1, it may be that large, heavy haul trucks control the design of horizontal curves rather than the smaller, faster pickup trucks (utes) and cars.

If a truck has a high center of gravity, and if its track width—the distance between the outside faces of the wheels at opposite ends of an axle—is relatively narrow, rolling over can be a danger on a tight horizontal curve (Figure 4.5). It may be that although the provision of a large enough radius of curvature and steep enough superelevation makes it safe for high speed pickup trucks (utes) and cars to negotiate a horizontal curve, the curve still may be dangerous for high, heavily loaded trucks with high centers of gravity. Roll-over of the heavy vehicle may govern the design, and make it necessary to specify a larger radius of curvature than could be tolerated by the lower vehicles.

Superelevation and minimum spiral parameter for low-volume roads, $e_{max} = 0.08$, normal cross slope = 0.04

Design speed, km/h	30		40		50		60		70		80		90		100	
Radius, m	e	A	e	A	e	A	e	A	e	A	e	A	e	A	e	A
7000	NC		NC		NC		NC		NC		NC		NC		NC	
5000	NC		NC		NC		NC		NC		NC		NC		0.040	530
4000	NC		NC		NC		NC		NC		NC		NC		0.040	480
3000	NC		NC		NC		NC		NC		0.040	350	0.040	390	0.040	410
2000	NC		NC		NC		0.040	260	0.040	275	0.040	300	0.040	300	0.040	335
1500	NC		NC		NC		0.040	225	0.040	255	0.040	250	0.040	270	0.040	290
1200	NC		NC		0.040	190	0.040	200	0.040	220	0.040	225	0.040	240	0.040	260
1000	NC		NC		0.040	170	0.040	175	0.040	200	0.040	220	0.043	225	0.047	240
900	NC		0.040	145	0.040	150	0.040	175	0.040	180	0.040	200	0.046	200	0.051	225
800	NC		0.040	135	0.040	150	0.040	160	0.040	175	0.042	175	0.049	200	0.055	210
700	0.040	110	0.040	125	0.040	140	0.040	150	0.040	165	0.046	175	0.053	185	0.061	200
600	0.040	100	0.049	120	0.040	125	0.040	140	0.042	150	0.050	165	0.058	175	0.067	190
500	0.040	90	0.040	100	0.040	120	0.040	125	0.048	140	0.056	150	0.064	160	0.073	190
400	0.040	80	0.040	90	0.040	100	0.045	115	0.054	125	0.063	135	0.071	160	0.080	190
350	0.040	75	0.040	90	0.040	100	0.049	110	0.058	120	0.067	125	0.075	160	0.080	190
300	0.040	70	0.040	80	0.042	90	0.053	100	0.063	120	0.072	125	0.080	160	Minimum R=390	
250	0.040	60	0.040	75	0.047	85	0.059	100	0.069	110	0.078	125	0.080	160		
220	0.040	60	0.040	70	0.051	80	0.062	95	0.073	110	0.080	125	Minimum R=300			
200	0.040	60	0.041	70	0.054	80	0.065	90	0.075	110	Minimum R=230					
180	0.040	60	0.044	65	0.057	75	0.068	90	0.078	110						
160	0.041	55	0.047	55	0.060	75	0.072	85	0.080	110						
140	0.044	55	0.051	65	0.064	70	0.076	85	Minimum R=170							
120	0.048	55	0.055	60	0.069	70	0.080	85								
100	0.053	50	0.061	55	0.074	65	0.080	85								
90	0.056	50	0.064	55	0.077	65	Minimum R=120									
80	0.060	50	0.067	55	0.080	65										
70	0.064	50	0.071	50	0.080	65										
60	0.068	45	0.075	50	Minimum R=80											
50	0.070	45	0.080	50												
35	0.080	40	Minimum R=50													
30	0.080	35														
	Minimum R=30															

$e_{max} = 0.08$

Normal cross slope = 0.04

Notes

e is superelevation
A is spiral parameter in metres
NC is normal cross section
Spiral length, $L = A^2$ / Radius
Spiral parameters are minimum and higher values should be used where possible
Spirals are desirable but not essential above the dashed line.

Figure 4.4 Typical horizontal curve design table. (Reproduced with the express written authority of the Transportation Association of Canada (TAC), Excerpted from the TAC publication *Geometric Design Guide for Canadian Roads*, Transportation Association of Canada, Ottawa, Ontario, Canada, 2011.)

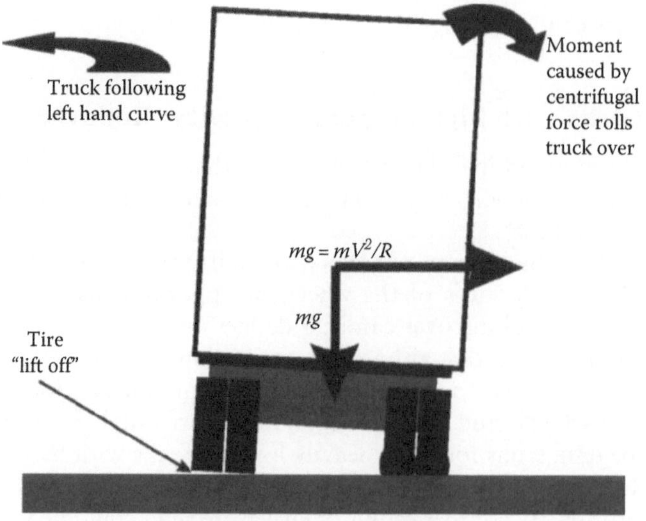

Figure 4.5 Truck on left-hand curve about to roll over.

Research work aimed specifically at this problem was carried out by El-Gindy and Woodrooffe [4.12], on a broad spectrum of heavy logging trucks. In general terms, their work determined through mathematical modeling of the behavior of the truck components in constant circular turns that roll-over would occur at side accelerations in the vicinity of $0.22g$ to $0.33g$ (where g is the acceleration due to gravity), depending upon the truck configuration. Given that the side acceleration is V^2/R, where V is the speed of the vehicle, and R is the radius of the horizontal curve (Figure 4.3), a table of minimum radii can be derived, based on the design speed of the road and some maximum acceptable side acceleration, set below the roll-over thresholds predicted by El-Gindy and Woodrooffe. This is provided in Table 4.4.

What the maximum acceptable side acceleration should be can be argued. Table 4.4 uses a maximum side acceleration of $0.15g$, that being about 2/3 of the minimum value El-Gindy and Woodrooffe found for the truck configurations they modeled [4.12]. There is a tradeoff: the higher the acceptable side acceleration taken, the tighter the horizontal curve may be, and therefore the less expensive the construction of the resultant roads will be in extreme cases. The designer will have the freedom to select a tight radius if that is desirable. However, the higher the acceptable side acceleration value is, the greater the risk of trucks rolling over.

Comparing the minimum allowable radii in Table 4.4 to the design values in Figure 4.4, it is clear that the roll-over criterion governs over the conventional side sliding criterion for the limiting cases (maximum superelevation, minimum radius). For lower rates of superelevation and correspondingly larger radii, the sliding criterion will govern.

4.7 SPIRAL TRANSITION CURVES*

To provide a smooth transition in curvature, from the tangent alignment of the road where the radius of curvature can be conceived as being infinite, to the circular curve, where the radius has been set at some value R, spiral transition curves are used. In Figure 4.6a, AB is the tangent alignment, BCD is the spiral curve, and D is where the circular curve begins.

Spiral transition curves also permit the development of superelevation, allowing the road cross section to change smoothly from the usual shape with normal cross fall on both sides of the road centerline, to the shape where the full superelevation has been developed. Usually, the road cross section is rotated about the centerline to work in these changes, as shown

* Spiral transition curves are rare in LVRs. Discussion of them is included here for completeness' sake.

Table 4.4 Minimum radius for roll-over criterion based on maximum lateral acceleration of 0.15g

Design speed (km/h)	*Minimum radius (m)*[a]	*Design speed (mph)*	*Minimum radius (ft)*[b]
30	45	20	180
40	85	25	280
50	130	30	400
60	190	35	550
70	255	45	900
80	335	50	1110
90	425	55	1350
100	525	60	1600

[a] Rounded to nearest 5 m.
[b] Rounded to nearest 10 ft.

in Figure 4.6b. The points A, B, C, and D are significant points along the curve as the superelevation is developed:

A: The last point on the alignment approaching a curve, where the road is tangent and the normal cross section prevails
B: The point where there is a change from tangent to spiral; point at which the outside half of the road is horizontal
C: Point where the road is still in spiral and the full road cross section is superelevated at e = normal crossfall for drainage
D: Point where there is a change from spiral to circular curve and e_{max} has been achieved

What should the length of the transition curve be? American [4.3], Canadian [4.5], British [4.6], Australian and New Zealand [4.9], and South African [4.10] practices all focus on the rate of roll rotation (recall Figure 4.1) and set the length of the transition great enough that the vehicle isn't snapped into the full superelevation of the circular curve. How the length is calculated and the charts are used varies from country to country, and reference

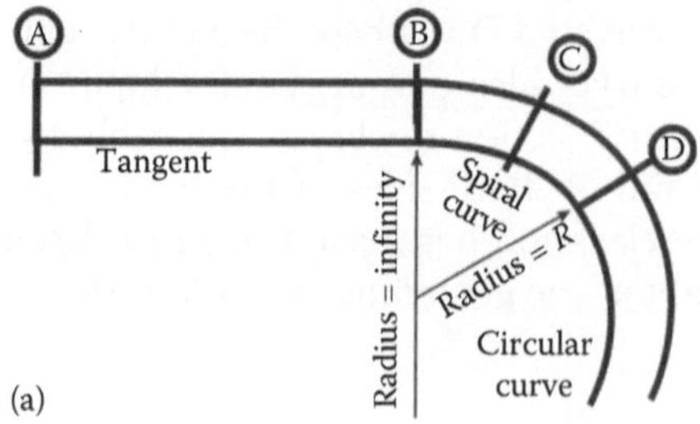

Figure 4.6 Horizontal curve: (a) plan view. *(Continued)*

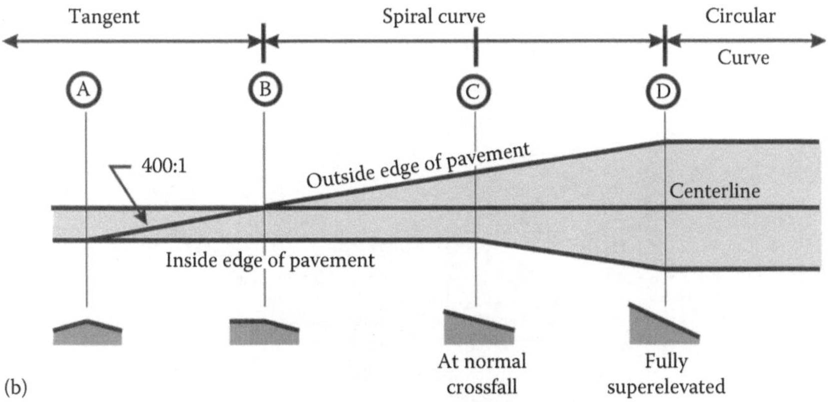

Figure 4.6 (Continued) Horizontal curve: (b) side view from inside the curve.

should be made to the relevant publication. All are in agreement, however, on the requirement to have developed the full superelevation by the time the vehicle reaches the circular curve (point D in Figure 4.6b).

4.8 LATERAL CLEARANCE AND CLEARED RIGHT OF WAY WIDTH

Lateral clearance as shown in Figure 4.7 is the distance C from the lane centerline, perpendicular to the road, to an obstacle to the line of sight. The distinction between stopping sight distance (measured along the road alignment) and the line of sight must be understood.

The problem is one of clearing obstacles to sight far enough back from the road that the line of sight to a hazard puts the vehicle at least the required stopping sight distance away from the hazard, measured along the lane centerline (Figure 4.7). The geometry of the problem leads to the following equation:

$$C = R \times \left[1 - \cos\left(28.6 \times \frac{S}{R}\right)\right] \tag{4.10}$$

where

- C is the required lateral clearance (see Figure 4.7)
- R is the circular curve radius
- S is the stopping sight distance
- *Note:* all units are either metres or feet, cos is in degrees.

Any obstacle at the side of the road blocking the driver's line of sight around the curve must be cut back a distance of at least C from the inside lane centerline.

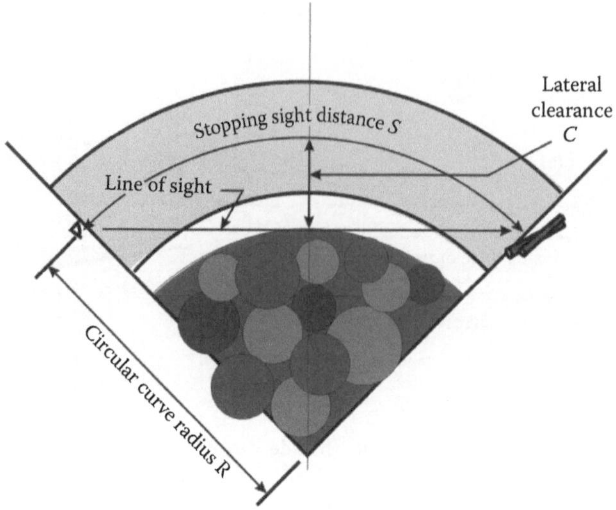

Figure 4.7 Lateral clearance around a horizontal curve.

4.9 RAILWAY CROSSINGS AT GRADE

The building of a railway crossing at grade can only be undertaken once the appropriate permits have been acquired: one does not just cross a railway with a road anywhere one pleases. None of the legal aspects of railway crossings will be discussed here; it is assumed that the necessary permission has indeed been obtained.

As shown in Figure 4.8, a diamond-shaped area has to be cleared on either side of the road and either side of the railway, to enable drivers approaching from either side to see trains traveling in either direction, well enough in advance of meeting a train to be able to stop. It is assumed the train will not slow down approaching the crossing—the vehicle on the road will have to stop. One side of a triangle is the stopping sight distance of the vehicle. Another side is the distance a train would cover, approaching the crossing, during the vehicle's stopping time. The lengths can be determined once the stopping sight distance and time are set, and once the design train speed is set. Table 4.5(a) and (b) shows the results of the calculations.

4.10 MINIMUM GRADIENT

The minimum longitudinal gradient is independent of the design vehicle (Table 4.1). Gradients are measured in units of vertical rise divided by horizontal run, usually expressed in percentages. The minimum gradient is set to cater to drainage requirements. In order to avoid the ponding of

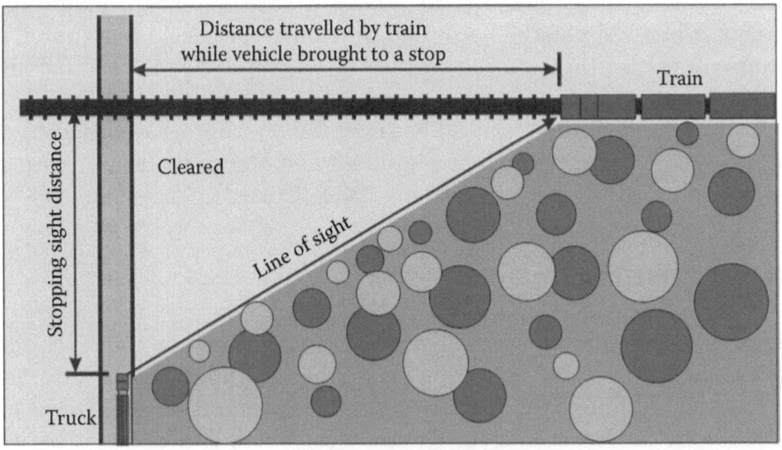

Figure 4.8 One quadrant of the cleared area at a railway crossing.

Table 4.5 Railway crossings
(a) metric units

Road			*Distance traveled by train at speed shown during time for vehicle to come to a stop (m)*			
Design speed (km/h)	*Coefficient of friction*	*Stopping sight distance (m)*[a]	*Train speed (km/h)*			
			30	*80*	*100*	*150*
60 or less	0.34	85	65	170	210	320
80	0.31	140	85	230	280	420
100	0.30	200	100	270	340	500

(b) U.S. customary units

Road			*Distance traveled by train at speed shown during time for vehicle to come to a stop (ft)*			
Design speed (mph)	*Coefficient of friction*	*Stopping sight distance (ft)*[b]	*Train speed (mph)*			
			20	*50*	*60*	*90*
35 or less	0.34	280	215	530	635	950
50	0.31	460	290	720	865	1295
60	0.30	660	340	855	1020	1535

[a] Road assumed to be level.
[b] Road assumed to be level.

water on the road surface, the road should have a nonzero longitudinal gradient wherever possible. Any length that is perfectly horizontal (for instance at the bottom of a sag vertical curve) should be kept to the absolute minimum.

The minimum gradient should be set at 0.5% or more. Roads with particularly rough surfaces should be built with an even greater minimum longitudinal gradient (or better surfacing material used).

4.11 MAXIMUM GRADIENT

Contrary to the minimum gradient, setting the maximum tolerable gradient is dependent on the vehicles with the highest mass/power ratio using the road. In addition, the maximum gradient should be set after a consideration of the economics of the project. There is a tradeoff between the cost of building the road, and the cost of operating the vehicles on it; extra money spent flattening grades will be returned in reduced vehicle operating costs, and often the operating costs far outweigh the original construction cost.

The values shown in Table 4.6 are drawn from a survey of the existing low volume roads standards of 63 countries made in 1975 by the World Bank [4.13]. The maximum gradients recommended in Table 4.6 ignore design speed, but reflect the expected overall geometric standard in the traffic volume. Note that the standards are relaxed for rougher topography, to make it possible to accommodate the topography economically.

4.12 CRITICAL LENGTH OF GRADE

The performance of a heavy vehicle while climbing a grade is very sensitive to the gradient. Therefore, while steep gradients can be accommodated in extreme cases, their lengths must be kept to a minimum if truck

Table 4.6 Maximum gradients (%)

	Class I	*Class II*
Topography	*ADT[a] = 0–50 veh/day*	*ADT[a] = 50–400 veh/day*
Flat	6	6
Rolling	8	7
Mountainous	10	9

Source: Cron, F.W., A review of highway design practices in developing countries. Reproduced as Text 5 in: [4.4] TRB. 1978. Compendium 1, geometric design standards for low volume roads, Transportation Research Board, National Research Council, National Academy of Sciences, Washington, DC, 1975.

[a] Average daily traffic.

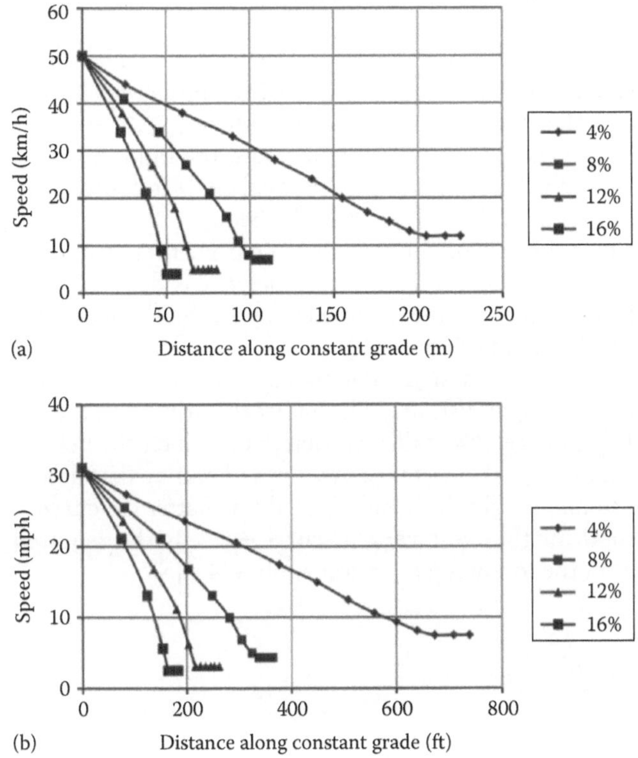

Figure 4.9 Predicted truck grade climbing behavior, mass/power ratio = 400 g/W (3 HP/ton): (a) metric units, (b) U.S. customary units.

performance is not to be adversely affected. Figure 4.9 shows the predicted variation in speed of a typical heavy vehicle while climbing constant grades. The truck loses speed rapidly on the grades, and eventually reaches an equilibrium speed, in most cases grindingly slow, where the power required to climb the grade is just balanced by the power output by the truck. It is clear from the figure that short stretches of what might be considered relatively moderate gradients have a dramatic impact on the performance of heavy vehicles.

On LVR, short stretches (up to 60 m or 200 ft) can accommodate gradients up to 20%, on paved roads with hard shoulders and full paving to drain edge, to minimize earthworks in unstable geology. On unpaved gravel or earth roads, stretches with gradients greater than 6% should be paved, stabilized, or reinforced with paving blocks to prevent erosion. The alternative to flattening grades is to use full-width paving from drain edge to the property line, with proper provision of side drainage.

4.13 CREST VERTICAL CURVES

It is customary to design with parabolic vertical curves, because they result in alignments comfortable to drive. One of the properties of parabolic vertical curves is that the rate of change of gradient (pitch—recall Figure 4.1) is constant throughout the curve. As a result, the sight distance offered to the driver of a vehicle negotiating the vertical curve is constant. In addition, parabolic curves are easy to lay out in the field.

The fact that the rate of change of gradient for these parabolic curves is constant is used to design them. A parameter K is defined, where K is the horizontal length of the curve, L, divided by A, the change in gradient over the length of the curve (Figure 4.10).

The requirement is to design a flat-enough vertical curve to give the driver sufficient stopping sight distance. The higher the driver sits, the farther down the road they can see. The taller an object for which the vehicle will have to be stopped, the greater distance away it first can be detected. Finally, the greater the stopping sight distance required, the flatter the curve must be.

If the length of the crest vertical curve exceeds the required stopping sight distance, the following equation applies [4.5]:

$$K = \frac{S^2}{200\,(\sqrt{h_1} + \sqrt{h_2})^2} \tag{4.11}$$

where

$K = L/A$ (m/%, or ft/%)

S is the required stopping sight distance (m or ft)

h_1 is the distance from the road surface to the level of the driver's eye (m or ft)

h_2 is the height of object on the road for which the vehicle must be stopped (m or ft)

L is the length of the crest vertical curve (m or ft)

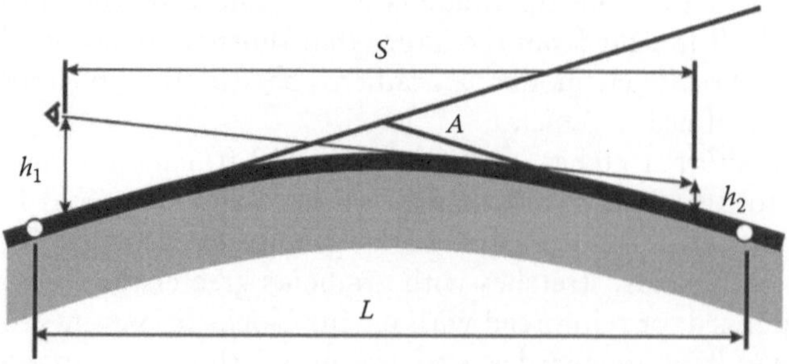

Figure 4.10 Crest vertical curve geometry, length of vertical curve > stopping sight distance.

A is the algebraic difference in gradients, between the entering tangent and the exiting tangent (i.e., the magnitude difference, respecting the signs of the gradients) (%)

Note: all length units are either metres or feet.

If the length of the crest vertical curve is less than the required stopping sight distance, then the following equation applies [4.5]:

$$K = \frac{2S}{A} - \frac{200\,(\sqrt{h_1} + \sqrt{h_2})^2}{A^2} \tag{4.12}$$

Inspecting the equation, it is apparent that the closer to the road surface the operator sits (the smaller h_1 is), the greater K must be. Also, the smaller the object for which it is assumed the vehicle must be stopped (the smaller h_2 is), the greater K must be. The greater K must be, the flatter the crest vertical curve.

Therefore, for this element of geometric design, the small, fast vehicles will govern, because h_1 is smaller and the stopping sight S is probably larger given the higher speeds (see discussion in Sections 4.2.1 and 4.3). Conventional practice [4.5,4.11] takes h_1 to be 1.05 m or 3.5 ft [4.3]. Object height h_2 is usually assumed to be 0.6 m or 2 ft.

4.14 SAG VERTICAL CURVES

The situation for sag vertical curves (Figure 4.11) is similar to that of crest vertical curves, but a difference is that for this component of geometric design, night conditions usually control. It is a matter of setting the K value so that the curve is flat enough that the area of roadway lit by the headlights of the vehicle extends far enough that adequate stopping sight distance is provided.

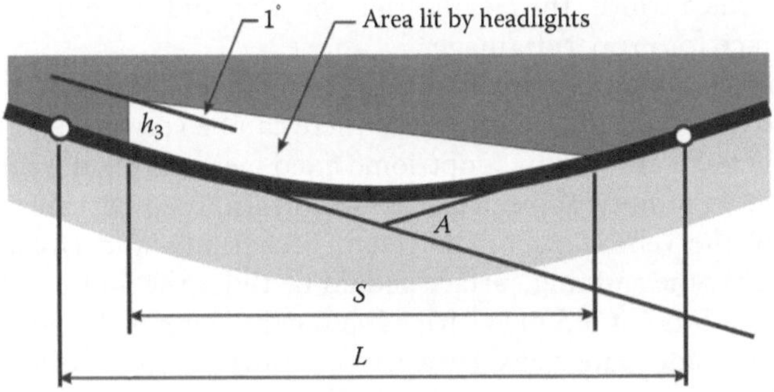

Figure 4.11 Sag vertical curve geometry, length of vertical curve > stopping sight distance.

If the length of sag vertical curve exceeds the stopping sight distance, then the following equation applies [4.5]:

$$K = \frac{S^2}{200(h_3 + S\tan\alpha)} \tag{4.13}$$

where
- $K = L/A$ (m/% or ft/%)
- S is the required stopping sight distance (m or ft)
- h_3 is the distance from the road surface to the level of the vehicle's headlights (m or ft)
- α is the angle of the headlight beam above the axis of the vehicle (degrees)
- L is the length of the crest vertical curve (m or ft)
- A is the algebraic difference in gradients, between the entering tangent and the exiting tangent (i.e., the magnitude difference, respecting the signs of the gradients) (%)

Note: all length units are either metres or feet.

If the length of the sag vertical curve is less than the stopping sight distance, then the following equation applies [4.5]:

$$K = \frac{2S}{A} - \frac{200(h_3 + S\tan\alpha)}{A^2} \tag{4.14}$$

Again, it seems most likely that the small, fast vehicles will govern for this element of geometric design. Generally h_3 is taken to be 0.6 m (2 ft), and α to be 1°.

4.15 CROSS-SECTIONAL ELEMENTS

The cross-sectional elements of interest to the designers of LVR include the running surface width, the cleared right of way, and the normal cross fall of the surface for proper drainage.

Conventional practice usually dictates some fixed lane width. However, for haul roads, it must be remembered there can be very large, often "oversized" vehicles. Rather than adopt some fixed measure for the running surface width, it is more appropriate to set the running surface width as a function of the vehicle width. Paterson, McFarlane, and Dohaney [4.14] suggested that the running surface should be twice the width of the design vehicle plus a total of 3.5 m (11 ft), as shown in Figure 4.12, to provide for shoulders and adequate space between opposing vehicles. Mining practice is to set the running surface width equal to four or five times the width of the design truck.

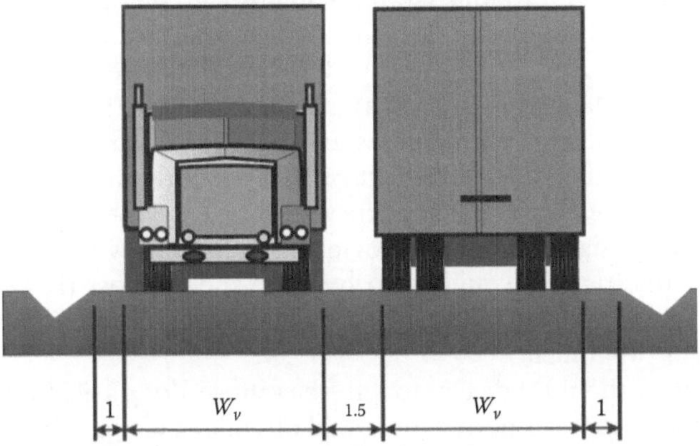

Figure 4.12 Road width calculation (metres: note 1 m = 3.3 ft).

The normal crossfall—the lateral gradient of the surface of the road—should be considered as one of the cross-sectional elements of geometric design. It should be set steep enough to ensure lateral drainage of surface water, but not so steep as to make steering of the vehicles difficult. For sealed surfaces, 2% normal cross fall is appropriate, but a value of 4% is more appropriate for the rough gravel and crushed rock surfaces used on access roads. A gravel road constructed with a 4% cross fall may have that reduced by traffic and grading, impairing surface drainage. The 4% cross fall must be carefully maintained.

4.16 ONE-LANE, TWO-WAY ROADS

The design recommendations presented in this chapter have assumed a two-lane, two-way road is being designed. However, it may be that the road has a single lane with passing places at intervals, presenting different conditions for some of the geometric design elements.

For these roads, the stopping sight distance should be doubled, because rather than a vehicle approaching a stationary object, the most hazardous condition on single-lane, two-way roads occur when two opposing vehicles approach one another. The closing speed is double the design speed. Doubling the stopping sight distance will in turn have an effect on the calculations for side clearance and cleared right of way width, and crest vertical curves.

4.17 GENERAL CONTROLS

A number of other aspects of geometric design should be mentioned, now that the specific design controls have been outlined. Generally, this is a matter of attempting to preserve the "flying vehicle" concept discussed at the beginning of the chapter.

It is most important to understand that although the specific elements of geometric design have been taken in isolation and discussed separately, it is necessary to design the alignment's horizontal and vertical components simultaneously. Dangerous situations sometimes can be inadvertently built into the alignment, if vertical curves and horizontal curves are separated. An example of this is the placement of a horizontal curve at the top of a long vertical curve. The horizontal curve will be encountered abruptly, and even if sufficient sight distance is provided, its abruptness still will present a hazardous condition. It would be far better to superimpose the vertical and horizontal curves so that the vehicle is taken through both simultaneously in a sweeping alignment.

The designer should endeavor to take the vehicle along a gentle alignment, with no sudden changes of direction, and no sudden changes in standard of geometric design. For example, a tight horizontal curve placed after a long alignment of flat horizontal curves and long tangents—where the operator has become accustomed to a high standard of geometric design—presents a dangerous situation. "Broken backed" horizontal alignments, where tight horizontal curves are separated by short tangent sections, are dangerous, causing vehicles with high centers of gravity to sway unduly. Closely spaced reversed horizontal curves are to be avoided for the same reason.

"Roller coaster" vertical alignments, where cut and fill operations have been kept to the minimum resulting in closely spaced crest and sag vertical curves, should be avoided. The resultant vertical alignment is abrupt and uncomfortable, and may create hazardous blind spots in the hollows.

Gradients should be reduced through intersections, to make it easier for vehicles to stop and maneuver. The sight lines for vehicles stopped to check traffic are also improved. The sites of intersections where one road meets another with high superelevation should be moved for this reason. Skewed intersections which meet at significantly less than 90° and intersections on curves should be avoided because of their poor sight lines. Recall too that "intersections" includes intersections of roads with railways (Section 4.9).

REFERENCES

4.1 IRC. 2015. *Catalogue of Indian Road Congress Publications.* http://www.irc.org.in/ENU/Documents/listpubl%20may2011_pdf.pdf, visited April 7, 2015.

4.2 PMGSY. 2015. Ministry of Rural Development, Government of India. http://pmgsy.nic.in/, visited April 7, 2015.

4.3 AASHTO. 2001. *Guidelines for Geometric Design of Very Low-Volume Local Roads (ADT ≤ 400).* Washington, DC: American Association of State Highway and Transportation Officials (AASHTO), 72pp.

4.4 TRB. 1978. *Compendium 1, Geometric Design Standards for Low Volume Roads.* Washington, DC: Transportation Research Board, National Research Council, National Academy of Sciences.

4.5 TAC. 2011. *Geometric Design Guide for Canadian Roads*. Ottawa, Ontario, Canada: Transportation Association of Canada.
4.6 TRRL. 1988. *Overseas Roadnote 6: A Guide to Geometric Design*. Crowthorne, U.K.: TRRL, 29pp.
4.7 Austroads. 2010. *Guide to Road Design—Part 3: Geometric Design*. Sydney, New South Wales, Australia: Australian Road Research Boards (ARRB).
4.8 Giummarra, G. 2001. *Road Classifications, Geometric Designs and Maintenance Standards for Low-Volume Roads*. Sydney, New South Wales, Australia: ARRB Transport Research Ltd.
4.9 ARRB (Giummarra, ed.). 2000. *Unsealed Roads Manual, Guidelines to Good Practice*. Vermont South, Victoria, Australia: ARRB Transport Research Ltd.
4.10 South African DoT. 1993. Geometric design of rural roads, Technical Recommendations for Highways TRH17. Pretoria, South Africa: Department of Transport, 120pp.
4.11 RTAC. 1986. *Manual of Geometric Design Standards for Canadian Roads*. Ottawa, Ontario, Canada: Roads and Transportation Association of Canada (RTAC).
4.12 El-Gindy, M. and Woodrooffe, J.H.F. 1990. Study of roll-over threshold and directional stability of log hauling trucks. Ottawa, Ontario, Canada: Division of Mechanical Engineering, National Research Council of Canada. Technical Report TR-VDL-002 NRCC No. 31274, 113pp.
4.13 Cron, F.W. 1975. A review of highway design practices in developing countries. Reproduced as Text 5 in: [4.4] TRB. 1978. Compendium 1, geometric design standards for low volume roads. Washington, DC: Transportation Research Board, National Research Council, National Academy of Sciences.
4.14 Paterson, W.G., McFarlane, H.W., and Dohaney, W.J. 1969. A proposed forest roads classification system. Woodlands Report No. 20. Montreal, Canada: Pulp and Paper Research Institute of Canada, 47pp.

Chapter 5

Pavement concepts

5.1 INTRODUCTION

While Chapter 7 covers pavement design, the basic concepts of pavements (this chapter), and an understanding of pavement materials (the next chapter) need to be discussed first. This chapter covers the purposes of pavements, the components of road cross sections, common pavement structures, and the load-carrying mechanisms the different pavement types develop.

5.2 PURPOSES OF PAVEMENTS

Pavements must serve multiple purposes. Layers of pavement materials are placed on the native soil—the subgrade—to protect it from the stresses imposed by vehicles that it cannot withstand itself. In fact, each layer of the pavement protects the layer beneath it, thus the strongest layers of the pavement are usually found at the top of the pavement structure. Pavements are designed to provide smooth and safe running surfaces for vehicles. The smoother the surface, the less wear and tear on the vehicles using it, and the faster the possible running speed. Pavements enhance the resistance to the scuffing and abrasion imposed by tires, particularly where vehicles impose severe shear stresses in areas such as intersections, sharp curves in the road, or parking areas. Safety is enhanced by high vehicle skid resistance, particularly in wet weather. While the pavement should be smooth on the macro scale for comfort, the texture of the surface and even the individual particles should be rough enough to generate sufficient wet skid resistance.

Well-designed pavements offer low rolling resistance to the tires of the vehicles using them. This is especially important for industrial haul roads. The stiffer the pavement structure, the lower the rolling resistance and therefore fuel consumption, and as a consequence, the lower the cost of operating the vehicles using the road.

While satisfying all these requirements, well-designed pavements make optimum use of the materials needed to construct them, striking the balance between pavement capital cost, design life, maintenance requirements,

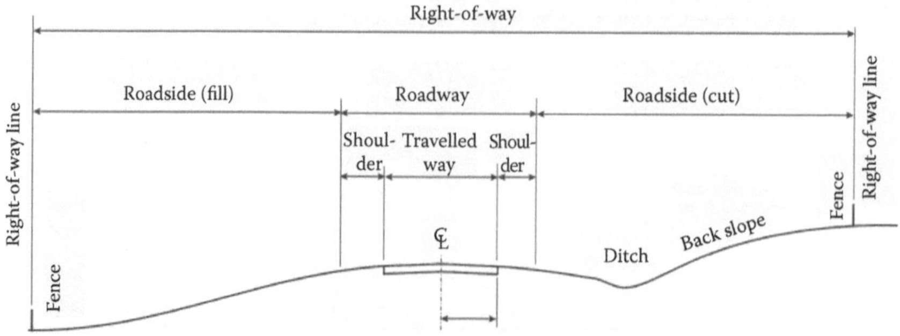

Figure 5.1 Road cross section terminology, North America.

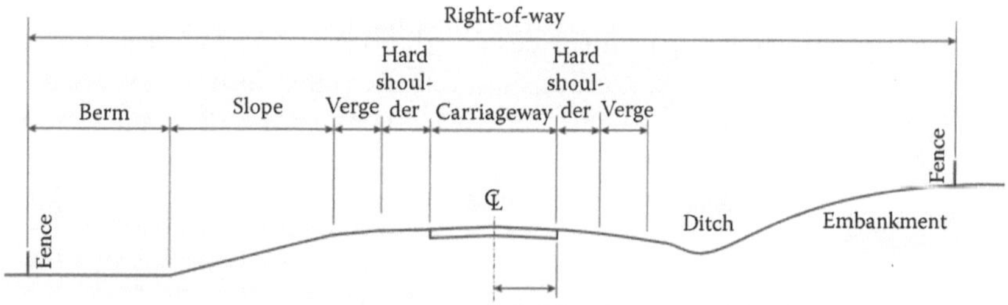

Figure 5.2 Road cross section terminology, Commonwealth countries except Canada.

and vehicle operating costs. Recently, more attention is being paid to the use of aggregates, with designers attempting to minimize the use of what is ultimately a finite resource. Interest in the use of marginal materials and recycled materials is rising.

5.3 ROAD CROSS SECTION COMPONENTS

It is vitally important that designers, constructors, and maintainers speak the same language when discussing the cross-sectional components of their pavements. Specifications and drawings must use consistent and recognized terminology. Unfortunately, the terminology is not consistent around the world. The terminologies prevalent inside and outside North America are indicated in Figures 5.1 and 5.2.

5.4 PAVEMENT STRUCTURES

Pavements are distinguished as *rigid* or *flexible*. Rigid pavements are constructed of Portland cement concrete. Such pavements are often just simply called "concrete," or, erroneously, "cement." Flexible pavements are essentially all other pavements: asphalt, bituminous surface treated, crushed rock or crushed gravel, or compacted soil. The layers in typical rigid and flexible pavements and their load-carrying mechanisms are different.

5.4.1 Rigid pavements

The surface layer of a rigid pavement is a concrete slab, which can be constructed in a number of ways, including

- *Plain, jointed, unreinforced*
- *Jointed and reinforced*, with or without dowels across the joints
- *Continuously reinforced*

Beneath the slab the pavement is built with a granular *base* (Figure 5.3a). The base provides drainage for water that penetrates the slab, as well as protection from frost damage. It also serves as a platform for construction of the slab.

Under the base is the *subgrade*, the native soil at the site (or the top of embankment fill if it exists). During construction, the subgrade should be proof-rolled before the pavement structure is built on it, and any soft, loose, wet, organic, or otherwise weak material subexcavated and replaced with clean, well-compacted base material fill.

5.4.2 Flexible pavements

Flexible pavements have a bituminous, aggregate, or soil surface on which the vehicles travel (Figure 5.3b). If the surface is asphalt, there may be more than one layer, with the *surface course* or *wearing course* underlain by one or more asphalt layers called the *binder(s)* or *base course(s)*. The surface course may be a bituminous surface treatment ("tar and chip," "chip seal"),

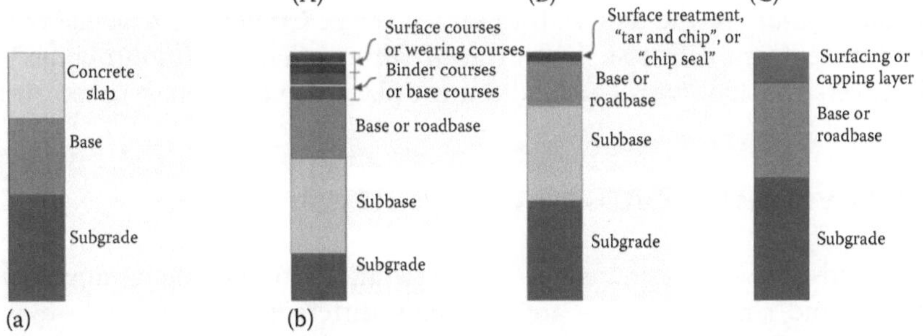

Figure 5.3 Pavement layers in (a) rigid pavements, and (b) flexible pavements ((A) asphalt, (B) surface-treated, (C) aggregate).

or it could be a granular layer consisting of crushed rock, crushed gravel, or gravel. Some designers of industrial haul roads call the surfacing course the *capping layer.*

Beneath the surface layer, flexible pavements usually have a granular *base* or *roadbase.* The base layer is usually designed to be well draining, so that water that penetrates the surface layer can be carried to the sides, out of the pavement structure. Water movement is enhanced by having all layers in the pavement crowned (cambered or constructed with a crossfall). Crossfalls of 2% and 4% are appropriate for asphalt and granular road surfaces, respectively. The granular road surfaces need a higher crossfall to overcome the greater impedance presented by the rough surface to the flow of water, because rutting may counteract drainage, and because grading operations tend to flatten the crossfall. It is a common error to specify too little crossfall for granular road surfaces—2% is inadequate.

Under the base layer the pavement may be designed with a *subbase* layer. The subbase is usually gravelly or sandy well-draining pit-run material. Beside drainage and protection of the subgrade beneath it, in cold regions the subbase is constructed of non-frost-susceptible material. This serves to increase the depth of non-frost-susceptible materials in the pavement structure, so the penetration of the frost front does not reach frost-susceptible soil and cause heaving.

The subbase is in turn underlain by the subgrade, the native soil at the site (or the top of embankment fill if it exists). During construction, the subgrade should be proof-rolled before the pavement structure is built on it, and any soft, loose, wet, organic, or otherwise weak material subexcavated and replaced with clean, well-compacted subbase material fill.

5.4.3 Variations

There are numerous variations beyond those listed here. For example, asphalt variations include full depth and deep strength pavements. The granular layers of any of the types of pavements can be treated with Portland cement, lime, fly ash, bitumen, or other chemicals or additives (as appropriate to the material) to stiffen them. Composite pavements are rigid pavements with the slab surfaced with asphalt. "White-topping" is the inverse of that, with Portland cement placed on an asphalt pavement.

5.5 PAVEMENT LOAD-CARRYING MECHANISMS

The load-carrying mechanisms—the resistances to bending, compressive, tensile, and shear stresses—are markedly different for rigid and flexible pavements. For all pavements, the goal is to reduce the stresses in the subgrade to a tolerable level, to avoid large strains and thus severe rutting or failure, but how the two types of pavement do this is different.

5.5.1 Rigid pavements

The slab of a rigid pavement has significant bending resistance, acting like a bending plate. Vertical deflection of the slab is negligible, and the vertical contact stress on the top of the base is widely spread (Figure 5.4a). In turn,

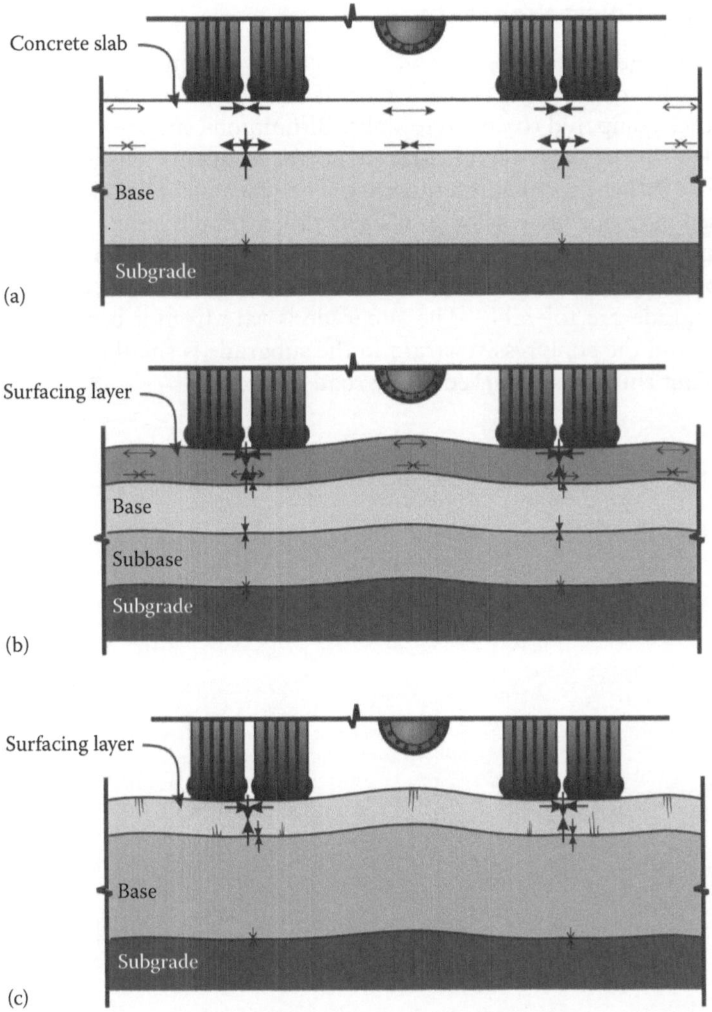

Figure 5.4 Load-carrying mechanisms for (a) rigid pavements, (b) flexible asphalt pavements, and (c) flexible granular pavements. Arrow direction and weight indicate compression or tension, and relative magnitude of the stresses. Note the bending stresses in the slab of the rigid pavement, the weaker bending stresses in the asphalt pavement, and the lack of tension in the granular pavement. Cracking in the granular pavement is shown on the top surface outside the dual wheels and in the bottom of the surfacing layer under the dual wheels. Note that for simplicity, the crossfall has not been shown.

the compressive stress transmitted by the base to the subgrade is dispersed and decreased even more, to the point where vertical displacement of the subgrade is negligible. Faulting at slab joints is avoided by specifying tongue and groove edges and/or dowels.

5.5.2 Flexible pavements

Unlike a rigid pavement, a flexible pavement exhibits detectable vertical elastic deformation under load. Asphalt surface layers have low bending resistance compared to concrete slabs. Bituminous surface treatments and granular surface layers have little or no resistance to tensile stresses and therefore little or no bending resistance. The pavement bends, resisting the imposed load in compression and shear rather than bending, the compressive and shear stresses diminishing with depth (Figure 5.4b and c). If the pavement layers are sufficiently thick and strong, the vertical stresses on the subgrade are tolerable. The subgrade is safe from a bearing capacity failure, and the compressive strain in the subgrade is small enough that no significant rutting is observed at the road surface.

Chapter 6

Pavement materials

6.1 INTRODUCTION

It is important to understand fully the materials used for the construction of LVR. The full array includes natural soils, aggregates, Portland cement, bituminous materials, and polymeric materials. While the polymeric materials (geosynthetics) will be covered in Chapter 12, the rest of the materials will be examined here. After discussing the basic phase diagram for soil and using it to define various gravimetric and volumetric terms, the materials and some of their tests will be covered in turn.

6.1.1 Sources of information

There is a wealth of information on road-building materials, including their manufacture or processing, testing, and specification. See Table 6.1 for some of the sources.

6.2 PHASE DIAGRAM

Pavement materials engineers and scientists use a particular vocabulary, with the terms stemming from the definitions of gravimetric (mass-based) and volumetric parameters derived from a diagram of material phases. Soils and aggregates have three phases: the solid phase, consisting of the mineral grains; the liquid phase, consisting of water (and possibly other liquids such as bitumen); and the gaseous phase, consisting of air and water vapor. Figure 6.1 shows each phase fused (lumped) together, with the left side showing the phases in terms of volume, and the right side showing them in terms of mass. Various key ratios can be derived as follows:

- The void ratio e and the porosity n are measures of the void volume of the soil, differing in the denominator of their ratios. Void ratio e is the more useful measure in consolidation (settlement) analyses because the denominator V_S remains constant as the soil is compressed.

Table 6.1 Sources of information on pavement materials

Author/agency	*Website[a]/publication*
AASHTO	http://www.transportation.org
ASTM	Annual book of ASTM Standards, Volume 04.08: Soil and rock, dimension stone, geosynthetics
Asphalt Institute	http://www.asphaltinstitute.org/
Asphalt Institute	The asphalt handbook, Manual Series No. 4 (MS-4), 7th edition
Asphalt Institute	Superpave, performance graded asphalt, binder specification and testing, Superpave Series No. 1 (SP-1)
Australian Road Research Board (ARRB)	http://www.arrb.com.au/Information-services/Publications.aspx
British Standards Institute	http://www.bsigroup.com/en-GB/
European standards	http://www.cen.eu/Pages/default.aspx
New Zealand Transport Agency	http://www.nzta.govt.nz/resources/index.html
Ontario Good Roads	http://applications.roadauthority.com/Standards/?Id=4EF3A115–7A19–4BC1–9BBF-689A6C8EC222
Ontario Ministry of Transportation	http://www.mto.gov.on.ca/english/pubs/index.shtml
South African National Roads Agency Ltd.	http://www.csir.co.za/Built_environment/Transport_Infrastructure/22RPF/05b.%20K%20Jenkins.pdf
Bowles, J.E.	Engineering properties of soils and their measurement [6.18]
Lambe, T.W.	Soil testing for engineers [6.19]
USDA Forest Service	http://www.fs.fed.us/publications/
U.S. Federal Lands Highway	http://flh.fhwa.dot.gov/resources/pse/specs
Wray, W.K.	Measuring engineering properties of soils [6.20]

[a] Websites visited November 3, 2014.

- The degree of saturation S indicates how much of the soil's void space is filled with water.
- The water content w is a key parameter in soil mechanics. Note that the denominator of its ratio is the mass of the soil solids, not the total mass.
- In pavement and geotechnical work, unit weight is usually a more useful parameter than density, making various calculations involving forces easier. Three unit weights are commonly defined: the bulk (moist but unsaturated) unit weight (in terms of water content or degree of saturation); the saturated unit weight, where all void space is filled with water; and the dry unit weight, where no water is present.

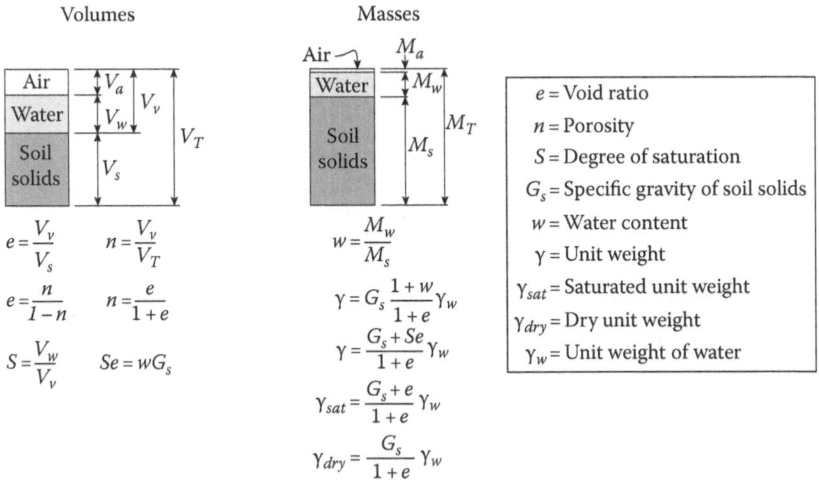

Figure 6.1 Volumetric and gravimetric soil phase diagrams, and the soil mechanics parameters derived from them.

6.3 SOIL TESTS

A handful of soil tests are routinely carried out, including tests for water content, plasticity, grain size distribution, and compaction.

6.3.1 Water content (AASHTO T 265, AS 1289.2.1.1, ASTM D 2216, BS 1377-2)

In the water content test, an appropriately sized sample of the soil or aggregate is first weighed in its moist state, then heated to 110°C until the dry weight is constant. The water content is the mass of the water driven off (moist mass–dry mass), divided by the dry mass.

6.3.2 Atterberg limits (AASHTO T 89 and T 90, AS 1289.3, ASTM D 4318, BS 1377-2)

Clays and silty clays possess plasticity—they deform at constant volume under pressure, then retain the new shape once the pressure is released. These soils can exist in any of four states, depending on the water content: solid, semisolid, plastic, and liquid (Figure 6.2). The Atterberg limits are the water contents at which the material is said to pass from one state to the next. The limits are somewhat arbitrary, but they are defined and tested for the same way around the world, and they are widely understood.

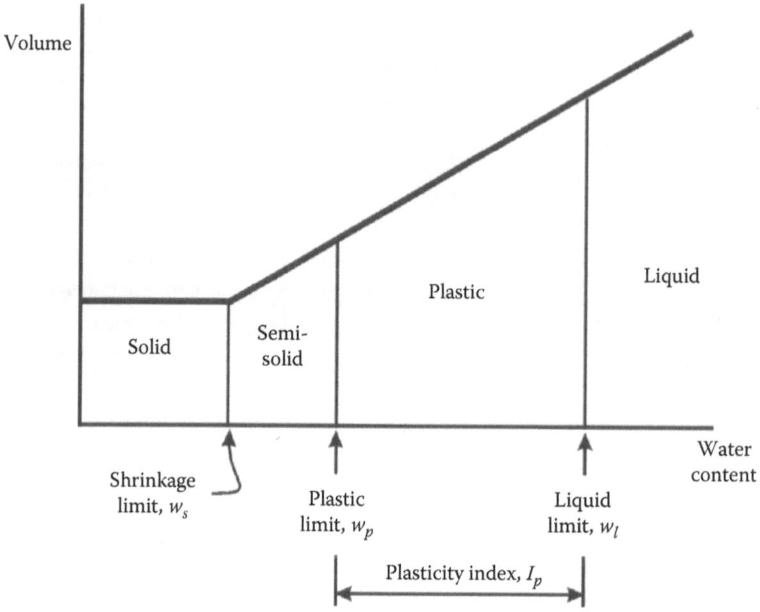

Figure 6.2 Changes of state of fine-grained soils, and the Atterberg limits.

The plastic limit is the boundary between the semisolid state and the plastic state. It is the water content at which threads of the soil can be rolled out by hand to a diameter of 3 mm (1/8 in.) on a frosted glass plate with the soil just beginning to crumble.

The liquid limit marks the change between plastic and liquid states and is determined either by using a dropping-cup apparatus (Figure 6.3) called the Casagrande apparatus, or a small weighted cone (a "fall cone," Figure 6.4) penetrating a sample of the soil. When using the Casagrande apparatus, a pat of soil is smeared into the bottom of the cup, and a V-shaped groove cut into it with a special tool. A cam mechanism lifts and drops the cup 10 mm (0.39 in.). The liquid limit is determined by counting the number of drops required to close the groove in the soil pat for a length of 10 mm (½ in.), and interpolating for 25 blows on a plot of water content vs. number of drops (Figure 6.5). The moisture content interpolated for 25 drops is taken as the liquid limit. If the fall-cone apparatus is used, soil samples are prepared at a range of water contents and smeared into individual cups. A 30° cone 35 mm (1.38 in.) long is allowed to penetrate each soil sample under its own weight for 5 s, and the penetration is recorded. Penetration vs. water content is plotted, and the water content at an interpolated penetration of 20 mm (0.79 in.) is the liquid limit.

Figure 6.3 Casagrande apparatus for determining liquid limit. (Photo courtesy of ELE International.)

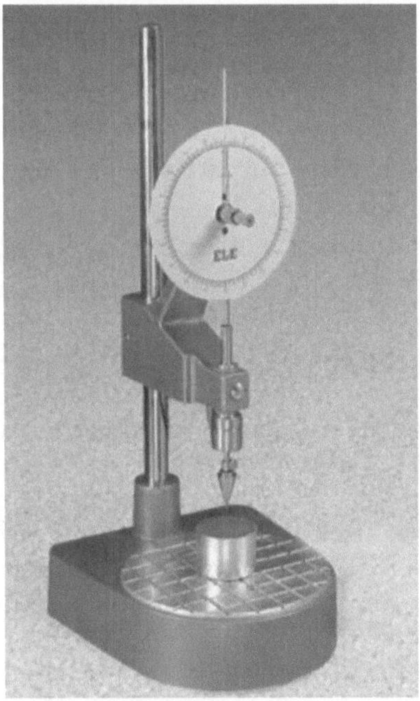

Figure 6.4 Fall cone for determining liquid limit. (Photo courtesy of ELE International.)

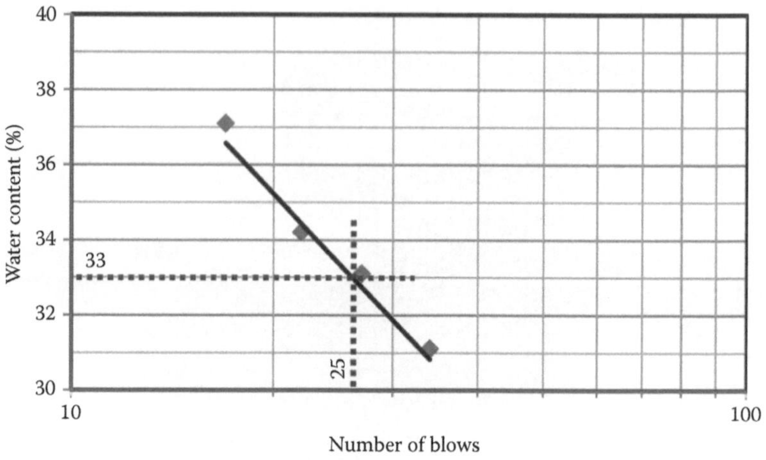

Figure 6.5 Example of graph to determine liquid limit from tests in the Casagrande apparatus. The liquid limit is the water content estimated for 25 drops on the best-fit line through the data.

An important parameter is the plasticity index of the soil—the difference between the liquid limit and plastic limit (Figure 6.2). The greater the difference, the greater the range of water content over which the soil will behave plastically.

6.3.3 Grain size distribution (AASHTO T 87 and T 88, AS 1289.3.6, ASTM D 421 and D 422, BS 1377-2)

The distribution of particle sizes is an important determinant of soil or aggregate behavior. The maximum particle size; the range of sizes; the proportions of clay, silt, sand, and gravel sizes; the amount of fines (grains which pass through the holes in a 75 µm [#200] sieve); and the uniformity of the particle sizes all play crucial roles in how a soil or aggregate behaves. To determine the grain size distribution, a representative sample is shaken through a nest of sieves (Figure 6.6), stacked with the opening sizes decreasing from the top sieve to the bottom sieve. Sieve hole sizes are either given in "hard" metric sizes, or by sieve number, based on the number of holes per inch (Table 6.2). The percentage of the sample by mass finer than a given sieve opening size is plotted against the opening size on a logarithmic scale (Figure 6.7).

It is good practice to use wash sieving techniques to obtain an accurate assessment of the fine content (finer than 75 µm [#200] size) of the soil or aggregate.

Figure 6.6 Stack of sieves and sieve shaker. (Photo courtesy of ELE International.)

Table 6.2 Sieve sizes (ASTM D422)

	Opening size	
Sieve identification	*(in.)*	*(mm)*
3 in.	3.00	76.2
2 in.	2.00	50.8
1½ in.	1.50	38.1
1 in.	1.00	25.4
3/4 in.	0.75	19.0
3/8 in.	0.375	9.52
#4	0.187	4.75
#10	0.0787	2.00
#20	0.0335	0.850
#30	0.0236	0.600
#40	0.0167	0.425
#50	0.0118	0.300
#60	0.00984	0.250
#100	0.00591	0.150
#200	0.00295	0.075

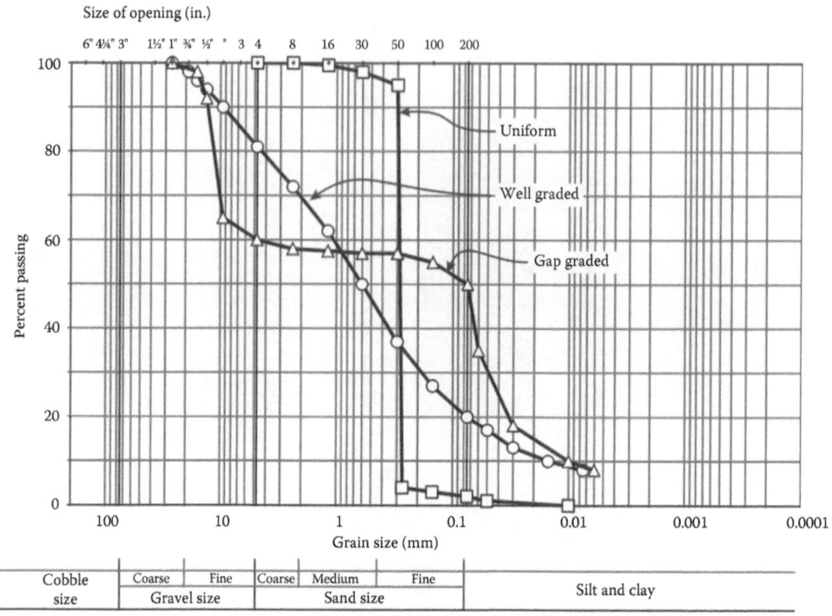

Figure 6.7 Typical grain size distribution curves.

For soils with an appreciable amount of material finer than 75 μm (#200 sieve), a sedimentation procedure is used. A 1 L slurry of soil is allowed to settle in a graduated cylinder, and the density of the slurry is measured periodically using a hydrometer. Stokes' law is used to determine the grain size distribution.

6.3.3.1 Features of the grain size distribution plot (C_u, C_c)

The character of the grain size distribution can be quantified by two indices, the coefficient of uniformity C_u and the coefficient of curvature, C_c (Equations 6.1 and 6.2).

$$C_u = \frac{D_{60}}{D_{10}} \tag{6.1}$$

where

- C_u is the coefficient of uniformity (unitless)
- D_{60} is the grain size for which 60% of the soil is finer (mm)
- D_{10} is the grain size for which 10% of the soil is finer (mm)

$$C_c = \frac{(D_{30})^2}{D_{60} \times D_{10}} \tag{6.2}$$

where

C_c is the coefficient of curvature (unitless)
D_{60} is the grain size for which 60% of the soil is finer (mm)
D_{30} is the grain size for which 30% of the soil is finer (mm)
D_{10} is the grain size for which 10% of the soil is finer (mm)

As shown on Figure 6.7, "uniformly graded" soils have essentially single-sized particles. "Well-graded" soils have a wide range of particle sizes. "Gap-graded" soils have a range of particle sizes missing from the soil.

6.3.4 Density

Soil or aggregate density is often required, for example, in compaction testing, slope stability analyses, or settlement analyses. There are a number of ways of determining density, either directly or indirectly, for example:

- *Directly with block samples.* A block of soil is excavated intact, weighed, sealed with wax, and submerged in a water bath to determine volume. Density is the mass divided by the total volume less the volume of wax used.
- *Directly using the sand replacement method.* A hole is dug in the soil with a soup spoon, and the soil retained for weighing. The volume of the hole is determined by filling it with sand of known density (Figure 6.8).
- *Directly using the balloon density method.* A hole is dug in the soil with a soup spoon through a plate with a hole in it and the soil retained for weighing. Volume is determined using a liquid-filled graduated cylinder whose lower end is covered by a rubber balloon (Figure 6.9). The cylinder is fitted over the plate before the hole is dug to get an initial reading on the cylinder's scale, the hole is dug, and a final volume reading taken on the graduated cylinder fitted over the plate again, the balloon taking up the shape of the hole. Volume is the difference in the readings taken on the graduated cylinder.
- *Indirectly using a nuclear apparatus* (Figure 6.10). The attenuation of radiation passing from the nuclear source in the tip of the unit's shallow probe back up to radiation sensors in the base of the apparatus is correlated to soil density.

The most common ways of measuring density are the balloon method and the nuclear method.

Figure 6.8 Apparatus to perform sand-replacement field density testing, which includes a cylinder filled with loose, dry sand of a known density. The cylinder has a valve in its bottom that permits the sand to flow out. To perform the test, (1) a flat surface is prepared on the soil and the flat plate placed on the soil, (2) a quantity of soil is dug through the plate's hole with a soup spoon and retained for weighing, (3) the cylinder filled with sand is placed on the plate and sand released into the hole, (4) the volume of the soil sample is determined by the volume of sand released into the hole and its density, and (5) the soil density is calculated by dividing the mass of the soil sample by the volume of the hole. (Photo courtesy of ELE International.)

6.3.5 Density in the laboratory (AASHTO T 99 and T 180, AS 1289.5, ASTM D 698 and D 1557, BS 1377-4)

If checking the compaction of fill material, knowing the *absolute* density of the compacted fill soil is relatively unhelpful. A standard is needed to compare the density being achieved in the fill to the maximum possible density. The Proctor moisture-density test provides the relevant data. Soil is compacted into a 100 mm (4 in.) diameter 115 mm (4.6 in.) tall mold using a drop hammer (Figure 6.11). Samples are prepared at a range of water contents. After compacting the soil into the mold, its density is determined. The dry density is plotted against water content for each sample (Figure 6.12) and the maximum dry density determined. The water content corresponding to this density—the standard Proctor maximum dry density

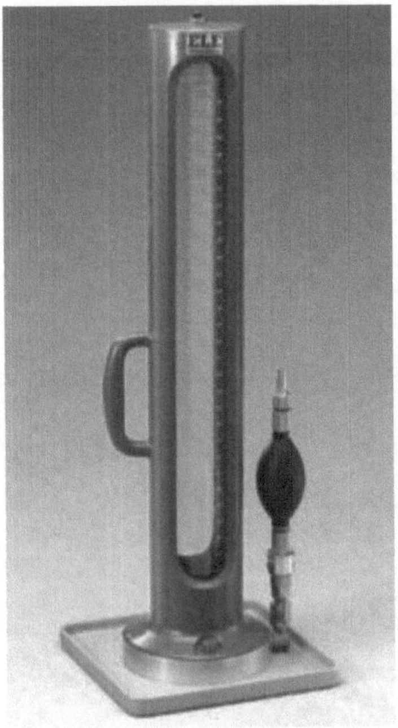

Figure 6.9 Apparatus to perform balloon test field density testing. The apparatus includes a graduated cylinder with a rubber membrane ("balloon") covering its bottom. The rubber squeeze ball and a valve on the side of the graduated cylinder permits the raising or lowering of the rubber membrane in the bottom of the cylinder by sucking air out of the top of the cylinder or readmitting air. To perform the test, (1) a flat surface is prepared on the soil and the flat plate placed on the soil, (2) the liquid-filled graduated cylinder is placed on the plate and an initial volume reading is taken, (3) the cylinder is removed and a quantity of soil is dug through the plate's hole with a soup spoon and retained for weighing, (4) the graduated cylinder is again placed on the plate, the rubber membrane in its bottom allowed to fill the hole, (5) a second volume reading taken on the graduated cylinder, and (6) the soil density is calculated by dividing the mass of the soil sample by the difference in the volume readings taken on the graduated cylinder. (Photo courtesy of ELE International.)

(SPMDD)—is the optimum water content (OMC) for compaction. Seldom do specifications call for less than at least 95% of standard Proctor maximum dry density (SPMDD), and densities measured in the field greater than Proctor density are possible.

6.4 SOIL CLASSIFICATION

Pavement materials need to be classified for all the reasons given in Section 3.1. Simple division of the soil into its fractions of clay, silt, sand, gravel, cobbles, and boulders leads to descriptions such as "silty clay,"

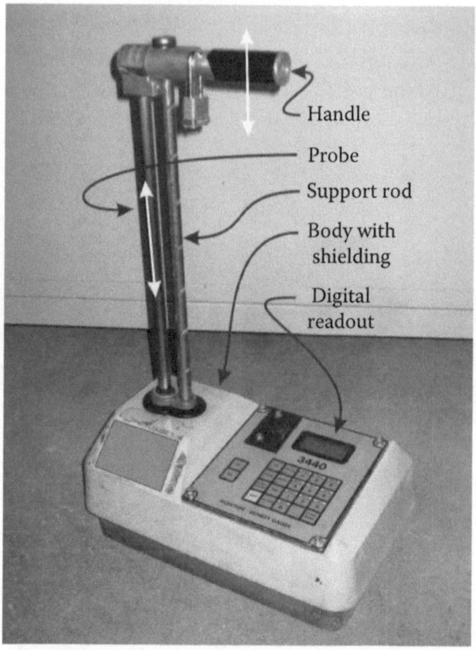

Figure 6.10 Apparatus to perform nuclear field density testing. The apparatus contains a radioactive source in the tip of a probe which can be raised through the body of the apparatus. In the photo, the probe is in the raised position. To perform the test, (1) a flat surface is prepared on the soil; (2) a rod is driven into the soil to create a hole; (3) the apparatus is placed over the hole and the probe lowered into the hole; (4) for a fixed time interval, a counter in the base of the apparatus measures the radiation emitted by the source and passing through the soil to the counter; and (5) the density of the soil is determined from calibration tables. (Photo with kind assistance of Darren Davis.)

which is a classification in essence. The divisions are somewhat arbitrary, but there are a few widely used schemes, such as the one shown in Table 6.3 [6.1].

There are other more comprehensive classification systems, and two in particular are commonly used, or form the basis for various agency and consultant in-house classification systems: the Unified Soil Classification System and the AASHTO classification system for subgrade material.

Both rely on aspects of the soil or aggregate's grain size distribution plot and if appropriate, the Atterberg limits of the fines in the soil. Lambe and Whitman [6.1, p. 35] and Wright and Paquette [6.2, pp. 424, 425] provide charts for the systems. Both systems of classification follow a left-to-right elimination scheme. Classification with the Unified Classification System yields a two-letter code, where the second letter indicates the major soil constituent, and the first letter the minor or

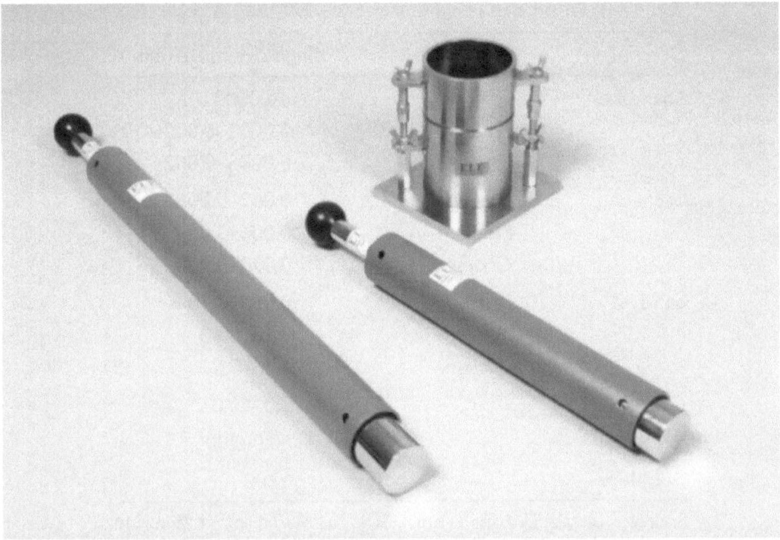

Figure 6.11 Apparatus for performing lab density tests. Soil is compacted into the mold using a drop hammer sliding in a tube. The standard Proctor test is carried out using the smaller hammer, the modified test using the larger hammer. Compactive energy for either test is controlled by specifying the weight of the hammer, the height of its drop (the length of the guide tube), the number of blows per layer of soil compacted into the mold, and the number of layers compacted into the mold. (Photo courtesy of ELE International.)

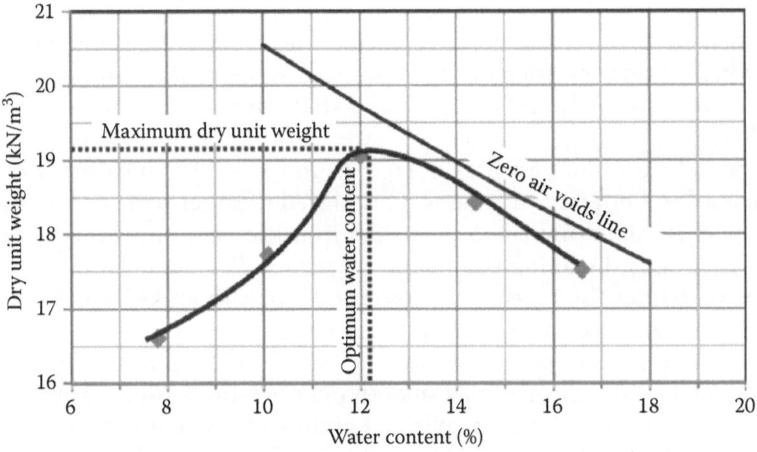

Figure 6.12 Example graph to determine standard Proctor maximum dry unit weight (or density) and optimum moisture content (OMC).

Table 6.3 Grain size ranges

Description		*Grain size range (mm)*
Clay sizes	Fine	<0.0002
	Medium	0.0002–0.0006
	Coarse	0.0006–0.002
Silt	Fine	0.002–0.006
	Medium	0.006–0.020
	Coarse	0.020–0.060
Sand	Fine	0.060–0.20
	Medium	0.20–0.60
	Coarse	0.60–2.0
Gravel		2.0–60.0
Cobbles		60.0–200
Boulders		>200

Source: Lambe, T.W. and Witman, R.V., *Soil Mechanics, S.I. Version*, John Wiley & Sons, Toronto, Ontario, Canada, 553pp., 1979.

modifying constituent. Using the AASHTO system results in a two- or three-character code, always beginning with A. The classification by the USCS method can then be used to determine the suitability of the soil for a given purpose [6.1, p. 37].

6.5 TESTS OF AGGREGATES

Routine tests performed on aggregates (sand and gravel, crushed gravel, crushed rock) for road building include tests for water content, grain size distribution, plasticity of the fines, Los Angeles abrasion, and micro-Deval abrasion. The first three tests listed have already been discussed earlier.

The L.A. abrasion test uses a large drum like that in a clothes dryer in which a few kilograms of aggregate is placed together with steel balls and tumbled. As the drum tumbles, a shelf in it lifts and drops the aggregate and steel balls, breaking the weaker particles. The "abrasion resistance" is the percentage by mass of the aggregate particles broken up. The test is rarely used and is only mentioned here for completeness.

A better test of abrasion resistance is the micro-Deval test [6.3]. The test uses a 5 L (5.3 qt.) drum with a smooth interior. An oven-dry sample of the aggregate is prepared with a specified grain size distribution and tumbled with a 5000 g (11 lb) charge of 9.5 mm (0.375 in.) diameter steel balls and 2 L (2.1 qt.) of water. The sample is dried and sieved again, and the micro-Deval abrasion loss calculated from the generation of fine material.

6.6 SPECIFICATION OF AGGREGATES

To ensure the properties of the materials used in the construction of a road coincide with what the pavement design assumes they will be, and for purposes of quality control, contract documents contain specifications for the aggregates to be used in construction. These are lists of tests and the required ranges of the results of the tests. A specification might include the grain size distribution "envelope," percentage crushed particles, and micro-Deval abrasion loss. Example specifications for crushed gravel base material and for crushed gravel road surfacing aggregate are given in Tables 6.4 and 6.5, respectively [6.4–6.7].

6.6.1 Surfacing aggregate for "unsealed" roads

Major* made the point that the evolution of "unsealed" roads has come full circle to an unfortunate state, where surfacing aggregate for them is concerned. Over the decades during his long association with national road construction in New Zealand, the specifications for road surfacing aggregates developed from rather open-graded aggregates, which performed poorly, to aggregates specifically chosen for their sealing characteristics. With their high fines contents, the surfacing aggregates that were developed actually sealed the so-called unsealed roads much as bituminous seals do in present times. The performance of roads constructed this way was good.

However, as traffic volumes grew in New Zealand over the decades of the last century, the "unsealed" roads were covered with bituminous surface treatments and asphalt. When covered, these *surfacing* aggregates made poor sealed road *bases* because the formerly desirable fines contents were too high to provide the drainage needed in their new role as base aggregates.

Base aggregate specifications with lower fines contents were developed. These perform well, providing both the needed drainage and sufficient stability.

Unfortunately, in present times, often the design engineer is unfamiliar with "unsealed" road requirements, and, when faced with the occasional "unsealed" road project, selects what they think is "the best" material, usually a well-graded 19 or 20 mm crushed rock *base* material with low fines content. The requirement for the surfacing aggregate is entirely misunderstood, and performance is poor.

For this reason, typical "unsealed" road surfacing aggregates—particularly the New Zealand specification [6.7]—are presented in Table 6.5, and their use is strongly advocated.

* Pers. comm.: Norman Major, New Zealand National Roads Board (retired).

Table 6.4 Example specifications for road base layer aggregate

		Province of Ontario, Canada [6.4]		*Province of Alberta, Canada [6.5]*		*Province of British Columbia, Canada [6.6]*	
Designation		OPSS granular A	OPSS granular O	Designation 2, base course aggregate		Base course aggregate	
		Road base under surfacing layer	Open-graded, free-draining base under surfacing layer	Class 20	Class 25	WGB: well-graded base, 25 mm	OGB: open-graded base, 25 mm
Source		Quarried bedrock; boulders, cobbles, gravel, sand, and fines from naturally occurring deposits	Quarried bedrock; cobbles or boulders retained on the 50 mm sieve				
Grain size distribution envelope							
Sieve size				**Percent passing**			
106 mm		n/a	n/a				
37.5 mm		n/a	100				
26.5 mm		100	95–100				
25.0 mm	1 in.				100	100	100
20.0 mm				100	82–97		
19.0 mm	3/4 in.	85–100	80–95			80–100	75–100
16.0 mm				84–94	70–94		
13.2 mm		65–90	60–80				
10.0 mm				63–86	52–79		
9.5 mm	3/8 in.	50–73	50–70			50–85	30–65
5.0 mm				40–67	35–64		
4.75 mm	#4	35–55	20–45			35–70	5–30
2.36 mm						25–50	0–10
1.25 mm				20–43	18–43		
1.18 mm		15–40	0–15			15–35	
630 μm				14–34	12–34		

(*Continued*)

Table 6.4 (*Continued*) Example specifications for road base layer aggregate

		Province of Ontario, Canada [6.4]		*Province of Alberta, Canada [6.5]*		*Province of British Columbia, Canada [6.6]*	
315 μm				9–26	8–26		
300 μm	#50	5–22	n/a			5–20	0–8
160 μm				5–18	5–18		
150 μm	#100	n/a	n/a				
80 μm				2–10	2–10		
75 μm	#200	2.0–8.0	0–5.0			0–5.0	0–5.0
Physical property requirements							
Percent crushed particles, % minimum		60	100				
Unconfined freeze–thaw, % maximum loss		n/a	15				
2 or more crushed faces, % minimum		n/a	85	60	60	50	50
Micro-Deval abrasion, coarse aggregate, % maximum loss		25	21			25	25
Micro-Deval abrasion, fine aggregate, % maximum loss		30	25				
LA abrasion, %				≤50	≤50		
Plasticity index, maximum		0	0	NP—6	NP—6	NP	NP

Sources: MTO, Ontario provincial standards for roads & public works: OPSS 1010 material specification for aggregates-base, subbase, select subgrade, and backfill material, Ministry of Transportation of Ontario, Toronto, Ontario, Canada. http://www.raqsb.mto.gov.on.ca/techpubs/ops.nsf/d37f5a16d8174ffa85256d130066857f/702bdff1eaaf4d9e85257b5c006080ac?OpenDocument, visited August 23, 2014; Alberta Infrastructure and Transportation, Standard specifications for highway construction, Professional Services Section, Alberta Infrastructure and Transportation, Edmonton, Alberta, Canada, p. 230, http://www.transportation.alberta.ca/Content/docType245/Production/Specs2007.pdf, visited September 2, 2014, 2007; British Columbia Ministry of Transportation and Infrastructure, Standard specifications for highway construction, BC Ministry of Transportation and Infrastructure, Victoria, British Columbia, Canada, http://www.th.gov.bc.ca/publications/const_maint/contract_serv/standard_ specs/SS202_Table_202-C-Aggregate_Gradations.pdf, visited September 2, 2014, 2012.

Table 6.5 Example specifications for granular road surface layer aggregate

		Province of Ontario, Canada [6.4]		*Province of Alberta, Canada [6.5]*		*Province of British Columbia, Canada [6.6]*	*New Zealand National Roads Board [6.7]*	
Designation		OPSS granular M	OPSS granular S	Designation 4, Gravel surfacing aggregate		Surfacing aggregate	20 mm maximum aggregate size	40 mm maximum aggregate size
		Unpaved road surfaces and for maintenance of unpaved shoulders	Dense-graded aggregates intended only for use as surfacing for low-volume unpaved roads with AADT <200 veh/day	Class 20	Class 40	HFSA	Granular road surfacing layer	Granular road surfacing layer
Source		Quarried bedrock; boulders, cobbles, gravel, sand, and fines from naturally occurring deposits	Quarried bedrock; boulders, cobbles, gravel, sand, and fines from naturally occurring deposits					
Grain size distribution envelope								
Sieve size				**Percent passing**				
106 mm		n/a	n/a					
40.0 mm					100			
37.5 mm		n/a	n/a					100
26.5 mm		n/a	100					
25.0 mm	1 in.					100		

(*Continued*)

Table 6.5 (*Continued*) Example specifications for granular road surface layer aggregate

		Province of Ontario, Canada [6.4]		*Province of Alberta, Canada [6.5]*		*Province of British Columbia, Canada [6.6]*	*New Zealand National Roads Board [6.7]*	
20.0 mm				100	55–90			
19.0 mm	3/4 in.	100	90–100			85–100	100	70–80
13.2 mm		75–95	75–100					
10.0 mm				35–77	25–72			
9.5 mm	3/8 in.	55–80	60–85			60–85	74–89	50–62
5.0 mm				15–55	8–55			
4.75 mm	#4	35–55	40–60			40–70	56–74	35–50
1.25 mm				0–30	0–30			
1.18 mm		15–40	20–40			20–50	36–51	22–36
300 µm	#50	5–22	11–25			10–30	25–36	16–26
150 µm	#150	n/a	n/a				17–28	12–19
80 µm				0–12	0–12			
75 µm	#200	2.0–8.0	9.0–15.0			5–15	10–21	8–16
Physical property requirements								
Percent crushed particles, % minimum		60	50	40	25			
Unconfined freeze–thaw, % maximum loss		n/a	n/a					
2 or more crushed faces, % minimum		n/a	n/a			50		

(*Continued*)

Table 6.5 (*Continued*) Example specifications for granular road surface layer aggregate

	Province of Ontario, Canada [6.4]		*Province of Alberta, Canada [6.5]*		*Province of British Columbia, Canada [6.6]*	*New Zealand National Roads Board [6.7]*	
Micro-Deval abrasion, course aggregate, % maximum loss	25	25			25		
Micro-Deval abrasion, fine aggregate, % maximum loss	30	30					
Micro-Deval abrasion, course aggregate, % maximum loss	25	25			25		
Micro-Deval abrasion, fine aggregate, % maximum loss	30	30					
LA abrasion, %			n/a	n/a		30–50	30–50
Plasticity index, maximum	0	0	NP–8	NP–8	≤6	6–15, Depending on clay content	6–15, Depending on clay content
Liquid limit						25–35	25–35
Lineal shrinkage, %						≤5	≤5
CBR						≥20	≥20

Sources: MTO, Ontario provincial standards for roads & public works: OPSS 1010 material specification for aggregates-base, subbase, select subgrade, and backfill material, Ministry of Transportation of Ontario, Toronto, Ontario, Canada. http://www.raqsb.mto.gov.on.ca/techpubs/ops.nsf/d37f5a16d8174ffa85256d13006 6857f/702bdff1eaaf4d9e85257b5c006080ac?OpenDocument, visited August 23, 2014; Alberta Infrastructure and Transportation, Standard specifications for highway construction, Professional Services Section, Alberta Infrastructure and Transportation, Edmonton, Alberta, Canada, p. 230. http://www.transportation.alberta.ca/Content/docType245/Production/Specs2007.pdf, visited September 2, 2014, 2007; British Columbia Ministry of Transportation and Infrastructure, Standard specifications for highway construction, BC Ministry of Transportation and Infrastructure, Victoria, British Columbia, Canada, http://www.th.gov.bc.ca/publications/const_maint/contract_serv/standard_ specs/SS202_Table_202-C-Aggregate_Gradations.pdf, visited September 2, 2014, 2012; Ferry, A.G., Unsealed Roads, a Manual of Repair and Maintenance for Pavements, Road Research Unit, New Zealand National Roads Board, Wellington, New Zealand, 1986.

6.7 SUBGRADE SOIL STABILIZATION MATERIALS

For some projects, it is desirable to treat the subgrade soil to increase its workability and compactibility. As a result, the soil's strength increases, causing an increase in bearing capacity, and its compressibility is decreased, causing a decrease in deflections and settlements. These are accomplished by mixing in an additive, such as lime or Portland cement, before compaction. This section covers lime and cement stabilization treatments, in brief summaries of the much more comprehensive discussions provided by Rodriguez et al. [6.8].

6.7.1 Stabilization with lime

6.7.1.1 Reaction

Stabilization with lime is thought to date back to the time of the Roman Empire. It is similar to stabilization with cement, although it works better with clayey soils and is less effective with coarse-grained soils than cement. Usually hydrated lime is used, as calcium hydroxide. Calcium carbonates do not react as effectively.

There are two chemical reactions. The first is instantaneous; soil particles capture calcium ions, which depress the diffuse double layer (a skin of water molecules bound to clay particles). The second reaction expands the double layer with the change in pH. Although it is a much slower reaction, taking days and weeks, it is the true cementing reaction.

6.7.1.2 Application

The reactions have little effect on organic soils or soils without the clay content needed to host the chemical reactions. Lime stabilization works well with plastic clays, which become more workable and easier to compact. Beside primary stabilization, lime treatment can be used as a pretreatment for cement stabilization because of these traits. Lime stabilization is not recommended for crushed rock or sands.

Strength gains can be significant; the unconfined compressive strength of silty clays can be doubled. Lime stabilization is effective for dosages of up to about 8% lime by weight or as much as 10% for high plastic clays. Curing takes days. The presence of organic content reduces effectiveness. Acidic water should be avoided.

In pavement applications, as a guide, typical lime contents are 3% for clayey gravel, 2%–4% for silty clay, and 3%–8% for high plasticity clay.

6.7.1.3 Construction

Typically, lime stabilization involves the following steps:

- Scarification
- Pulverization
- Spreading of the lime
  - Sprinkled evenly manually from bags or with a mechanical spreader
  - The application rate is usually determined from the results of a series of unconfined compression tests of soil–lime mixtures of different lime content
- Mixing
  - Using a mechanical mixer or by grader blading
- Addition of water
  - Water requirement is based on laboratory tests for the optimum water content for compaction
- Compaction
- Surface shaping
- Curing for at least 5 days
- Covering with a protective layer of paving before traffic is permitted

6.7.2 Stabilization with Portland cement

6.7.2.1 Reaction

The improvements in coarse soils derive from a chemical reaction between the cement and siliceous soil components. The reaction binds gravel, sands, and silt. In clayey soils, calcium hydrate is released as cement comes in contact with water, pulling water from the surface of clay particles. This reduces porosity and plasticity, and increases strength. As with lime stabilization, organics impede the reactions and for cement treatments and stabilization is unlikely to be effective if the organic content of the soil to be improved is more than 1% or 2%. Sulfates also weaken the cement.

6.7.2.2 Application

Cement is more often used to stabilize coarse-grained soils than fine-grained soils. Rodriguez et al. [6.8] relate the percentage of cement to be applied to a soil group index, which is dependent on the percentage of the soil finer than 0.075 mm (#200 sieve) and 0.5 mm (0.020 in.), the liquid limit, and the plasticity index. Recommended application rates range from about 3% to 8% by mass for GW, GP, GM, and SW soils, to 9%–15% for CH soils. The application rate is often determined by the results of a program of

Table 6.6 Guide to cement contents for cement-treated pavement materials

Soil type	*Typical percent cement by mass*	*Remarks*
Rock		
Crushed rock	0.5–2	Cement used to improve workability, to reduce soil's sensitivity to compaction water content, and to avoid deformation under heavy traffic
Sandy material		
Well-graded sandy clayey soils	2–4	
Well-graded sands	2–4	
Uniform sands	4–6	Compaction may be difficult and there is a risk of segregation of the cement
Clayey material		
Sandy clay	4–6	
Silty clay	6–8	
Clay	8–12	
Very active clay	12–15	Mixing may be very difficult and pretreatment with lime is usually beneficial
Organic material		
Organic soils	10–15	Pretreatment with lime or 2% $CaCl_2$ may improve cement's ability to set

Source: Rodriguez, A.R., del Castillo, H., and Sowers, G.F., *Soil Mechanics in Highway Engineering*, Trans Tech Publications, Clausthal-Zellerfeld, Germany, 843pp., 1988.

unconfined compression (UC) tests performed on samples with a range of cement contents. The recommended target UC strengths range from 0.35 to 1.0 MPa (50–150 psi) for subbases, 0.7–1.4 MPa (100–200 psi) for subbases or bases subjected to light traffic, and 1.4–5.5 MPa (200–800 psi) for bases subjected to heavy traffic. The strength is gained with unloaded curing over days, rather than hours.

The amount of water required usually depends on compaction requirements, with the optimum water content for compaction (OMC) the target. Strength partly depends on how well the treated soil is mixed; mixing is particularly difficult for soft clays. Table 6.6 shows a guide to the expected cement contents for paving materials.

6.7.2.3 Construction

The steps in constructing cement-treated pavement layers are similar to those for lime stabilization:

- Preparation of the soil—scarification and pulverization, and prewetting if necessary
  - Coarse-grained soils do not usually require scarification, whereas fine-grained soils do
  - Lumps should be less than 30 mm (1.25 in.) in size
  - The maximum particle size should be <50 mm (2 in.)
  - Proportioning and spreading of cement
  - Cement should be spread as evenly as possible
- Application of water
  - Plastic soils should be mixed at water contents less than the OMC
- Mixing
  - The more thorough the mixing, the higher the strength
  - If available, batch-plant mixing is the best
- Compaction
  - The usual requirements for compaction apply
    - Sheepsfoot rollers are recommended for cohesive soils, and rubber-tired or smooth-drum rollers for noncohesive soils
- Finishing, shaping
- Curing
  - The moisture content must be maintained to support the hydration of the cement
    - The treated soil should be covered with plastic or fabric continually wetted, or sprayed with a water-proofing compound

6.8 ASPHALT CEMENT

Many names are given to the sticky, black material used in pavements, including asphalt, asphalt cement, bitumen, and binder. The material is in turn an ingredient in asphalt concrete, hot mix asphalt, hot laid asphalt, bituminous surface treatments, tar* and chip, road tar, chip seal, and various slurries and seals.

ASTM defines asphalt as "*a dark brown to black cementitious material in which the predominating constituents are bitumens that occur in nature or are obtained in petroleum processing.*" Further, "*bitumen is a class of black or dark-coloured, solid, semi-solid, or viscous cementitious substances, natural or manufactured, composed principally of high molecular weight hydrocarbons, of which asphalts, tars, pitches, and asphaltites are typical*" [6.9]. Asphalt is present in most crude petroleums [6.9], and crude petroleum is the source of most asphalt cement used in paving.

To avoid confusion, the next sections will use the term "asphalt cement" as much as possible.

* Due to their hazardous nature, tars are not used in paving, although the names "tar and chip" and "road tar" persist.

6.8.1 Manufacture of asphalt cement

Most asphalt cement is derived from the petroleum refining process, where crude petroleum is distilled into its fractions, from lightest to heaviest being naptha, gasoline, kerosene, light burner oils, diesel oil, lubricating oils, and the residuum [6.9]. The heavy residuum is the source of the many grades of asphalt cement (Table 6.7).

Asphalt cement has three important characteristics: consistency, purity, and safety [6.9].

6.8.1.1 Consistency

Asphalt cements are thermoplastic, liquefying when heated. Several grades are identified according to their ranges of consistency. Historically, the penetration test (see Section 6.8.2.1) has been used to grade asphalt cement, with the grades 40–50, 60–70, 85–100, 120–150, and 200–300 being common. The designation is the range of acceptable penetration values. The softest is 200–300 pen, soft enough at room temperature that it can be indented with finger pressure. At room temperature, the hardest, 40–50, only allows a slight finger print to be made with firm pressure [6.9].

Consistency is better characterized by the viscosity of the asphalt cement. Viscosity is the resistance to shearing and is dependent on temperature. Therefore, when viscosity is measured, temperature must be specified. Two standard temperatures are used: 60°C (140°F) typifies the maximum temperatures expected in pavements in the United States and other regions where there is a significant amount of asphalt pavement, and 135°C (275°C) typifies the temperature of asphalt cements during mixing and laydown during construction.

6.8.1.2 Purity

Asphalt cement is almost entirely bitumen. Bitumen is entirely soluble in carbon disulfide [6.9]. Asphalt cement is almost entirely soluble, and impurities are inert. Normally, asphalt cement is free of water when it leaves the refinery. There is the chance that there will be water in truck tanks when it is transported to the road site. If water is present in the asphalt cement at temperatures above 110°C (230°F) the asphalt cement will foam, which is dangerous [6.9].

6.8.1.3 Safety

Foaming is a safety hazard. Specifications usually require no foaming below 175°C (350°F) [6.9]. If heated enough, asphalt cement releases vapors that will flash when exposed to flame. This temperature is the flash point. It is specified to be well above the temperatures in paving operations.

Table 6.7 Asphalt cement designations

Asphalt cements			*Emulsified asphalts*[a]		*Cutback asphalts*[b]	
Viscosity graded, original	*Viscosity graded, residue*	*Penetration graded*	*Anionic*	*Cationic*	*Medium curing*	*Slow curing*
AC-40	AR-16000	40–50	RS-1	CRS-1	MC-30	SC-70
AC-20	AR-8000	60–70	RS-2	CRS-2	MC-70	SC-250
AC-10	AR-4000	85–100	MS-1, HFMS-1	CMS-2	MC-250	SC-800
AC-5	AR-2000	120–150	MS-2, HFMS-2	CMS-2h	MC-800	SC-3000
AC-2.5	AR-1000	200–300	MS-2h, HFMS-2h	CSS-1	MC-3000	
			HFMS-2s	CSS-1h		
			SS-1			
			SS-1h			

Source: AI., *The Asphalt Handbook*, Manual Series No. 4 (MS-4), 7th edn., The Asphalt Institute, Lexington, KY, 788pp., 2010.

[a] Emulsified asphalt grades shown are AASHTO and ASTM grades and may not include all grades produced in any particular geographical area.

[b] Before considering the use of cutback asphalt cement, check applicable environmental regulations.

6.8.2 Tests of asphalt cement [6.9]

6.8.2.1 Penetration test

In the penetration test, asphalt cement is placed in a metal cup and held at 25°C (77°F). A needle of specified dimensions, weighted to 100 g is allowed to penetrate the sample for 5 s. The penetration in units of 0.1 mm over this time is that penetration value. For example, a penetration of 4.0 mm in 5 s gives a penetration of 40 pen.

6.8.2.2 Viscosity tests

The penetration test is empirical. A better characterization of the consistency of asphalt cement is its viscosity. Viscosity is used to grade asphalt cements. It is determined at 60°C (140°F) and 135°C (275°F). The maximum service temperature is expected to be no more than 60°C (140°F) and the mixing and pavement laydown temperature is expected to be no more than 135°C (275°F).

For the determination of viscosity at 60°C (140°F), a capillary tube viscometer is used. Vacuum is applied to pull the asphalt cement through a capillary tube. The time taken to flow between two marks on the tube is calibrated against viscosity. The water bath is used to keep the temperature constant at 60°C (140°F). The asphalt cement is graded according to its viscosity at 60°C (140°F) in Poises (P), where 1.0 P is 1.0 g/(cm·s). The SI metric units are Pa·s (=1.0 N·s/m^2), where 1.0 Pa·s = 10 P.

Asphalt cement is much more fluid at 135°C (275°F) and flows by gravity. Typically, a Zeitfuchs viscometer is used. The asphalt cement flows between two marks on the tube and the time taken is calibrated against viscosity. An oil bath is used to maintain the temperature at 135°C (275°F). The viscosity is reported in units of centistokes (1.0 cSt = 1.0 mm^2/s).

6.8.2.3 Flash point test

When heated enough, asphalt cement releases vapors which are combustible. The flash point is the temperature at which vapors will ignite in the presence of open flame.

6.8.2.4 Thin film oven test

The thin film oven test is actually a procedure, not a test. In the procedure, a film of approximately 10 mm (3/8 in.) is heated in a flat pan for 5 h in an oven at 163°C (325°F). Heating can age-harden asphalt cement. The procedure is to test the asphalt cement for penetration or viscosity before and after heating. The values after heating reflect any age-hardening that may have taken place.

6.8.2.5 Rolling thin film oven test

The rolling thin film oven test is similar to the thin film oven test but the apparatus and procedure are different. Asphalt cement is held in cylindrical glass bottles, and the bottles rolled horizontally within the oven. As it rolls within the bottles, the asphalt cement forms a thin film, continually exposing new surfaces to heating. The procedure is performed at 163°C (325°F) and penetration and/or viscosity tests are carried out after heating. A series of grades are based on the rolling thin film oven test: AR-1000, AR-2000, AR-4000, AR-8000, AR-16000, with a number indicating viscosity in Poises measured after the rolling thin film oven test, and AR signifying "aged residue."

6.8.2.6 Ductility test

In the ductility test, a briquette of asphalt cement is pulled horizontally at 50 mm (1.97 in.) per minute at a temperature of 25°C (77°F). The elongation in centimetres at breakage is the ductility.

6.8.3 Specification of asphalt cement

A list of the specified tests for asphalt cements is given in Table 6.8.

6.8.4 Cutback and emulsified asphalt cement

There are other ways to achieve the consistency of asphalt cement needed other than through the distillation of crude petroleum. It can either be "cutback," that is, mixed with lighter fractions of petroleum, or it can be constituted as an emulsion.

6.8.4.1 Cutbacks

In cutbacks, the asphalt cement is mixed with a petroleum solvent to alter its consistency:

- RC (rapid curing) cutbacks use a light diluent (solvent) such as gasoline or naptha
- MC (medium curing) cutbacks use a medium diluent such as kerosene
- SC (slow curing) cutbacks use oils of low volatility

However, it is no longer generally acceptable for environmental and safety reasons to use cut backs, as they release harmful vapors. Check the applicable environmental regulations if considering using one.

Table 6.8 Required tests for asphalt cement used in pavements

		Test method		*Asphalt cement specifications*[c]	
		AASHTO	*ASTM*	*Viscosity graded*	*Penetration graded*
Unaged asphalt cement	Viscosity at 60°C (140°F)	T 202	D 2171	√	
	Viscosity at 135°C (275°F)	T 201	D 2170	√	
	Penetration at 25°C (77°F)	T 49	D 5	√	√
	Flash point	T 48	D 92	√	√
	Solubility in trichloroethylene	T 44	D 2042	√	√
	Ductility at 25°C (77°F)	T 5	D 113		√
	Spot test[a]	T 102	—	√	√
Aged residue	Mass loss[b]	T 179	D 1754	√	√
	Viscosity at 60°C (140°F)	T 240	D 2872	√	
	Viscosity at 135°C (275°F)	T 202	D 2171	√	
	Penetration at 25°C (77°F)	T 201	D 2170	√	
	Ductility at 25°C (77°F)	T 49	D 5	√	√
		T 5	D 113	√	√

Source: Courtesy of AI, *The Asphalt Handbook*, Manual Series No. 4 (MS-4), 7th edn., The Asphalt Institute Inc., Lexington, KY, 788pp., 2010.

[a] Optional test in AASHTO specification.

[b] Mass loss is conducted in either TFO or RTFO (TFO listed first). Residue obtained after aging.

[c] Viscosity specifications are AASHTO M 226 and ASTM D 3381. Penetration specifications are AASHTO M 20 and ASTM D 946. Note that the viscosity specification includes three grading tables for which test requirements may differ from what is shown in the table.

6.8.4.2 Emulsions

To form an emulsion, asphalt cement is mechanically milled into tiny 5–10 μm (2×10^{-4} to 4×10^{-4} in.) globules and dispersed in water. This imparts the consistency needed during construction. The emulsion is said to "break" when eventually the water releases the globules onto the surface of the aggregates, cementing them, and the water evaporates. Emulsions may be anionic, cationic, or nonionic, having negatively charged, positively charged, or neutral globules. Selecting the correctly charged emulsion is required to make it compatible with the aggregate, whose surface may be positively or negatively charged. In the list of emulsions in Table 6.7, RS, MS, and SS indicate rapid, medium, and slow setting rates, and *h* indicates that a harder asphalt cement is used. High float asphalt cements, a type that results in thicker films on aggregates, are indicated by HF.

6.8.5 Performance-graded binders

It is possible to have different asphalt cements all within the accepted ranges for penetration and viscosity at 60°C (140°F) and 135°C (275°F), but still perform differently. It is preferable to be able to predict the performance of asphalt concrete in service based on physical and mechanical characteristics. In conjunction with the Superpave research program, a USD 50 million research project initiated in 1987 to find new tests and approaches, a new family of performance-graded binders was created. The term "binder" is intended to encompass both asphalt cements and modified asphalt cements. The selection of the grade depends on the climate in the geographical area, reflecting the asphalt cement temperatures during service. The grades are designated by the temperature ranges over which they are to perform satisfactorily, as shown in Table 6.9. For example, a PG 64-22 performance-graded binder is expected to perform satisfactorily over the range –22°C (–8°F) to 64°C (147°F).

The tests used to specify performance-graded binders are the following:

- The rolling thin film oven test
- The pressure aging vessel test
- The dynamic shear rheometer test
- The rotational viscometer test
- The bending beam rheometer test
- The direct tension test

See the Superpave manual [6.10] for test procedure details.

6.9 CHIP SEALS

An improvement on aggregate roads can be to seal them with chip seal ("tar and chip," "road tar"*). Chip seals are nonstructural layers placed on the pavement to provide a high fiction, low maintenance cost surface.

New Zealand has a long history of research and development on chip seals, which are widely used there [6.11]. Of the 94,000 km of roads in New Zealand, about 65% is paved, and the majority of the roads are chip sealed. Even high volume roads are chip sealed with some chip sealed roads having traffic volumes of AADT greater than 20,000 veh/day.†

To build chip seals, asphalt cement is sprayed on a compacted base, and single-sized, near-cubic stone chips are spread uniformly on the asphalt cement. The asphalt cement application rate, measured in litres per square

* Although for safety and environmental reasons, tars are no longer generally used and asphalt cement is used as the binder instead, the terms (erroneously) persist.

† Pers. comm. David Alabaster, NZ Transport Agency, December 9, 2014.

Table 6.9 Performance graded binders

	Low temperature maximum, °C						
High temperature minimum, °C	−10	−16	−22	−28	−34	−40	−46
46					PG 46-34	PG 46-40	PG 46-46
52	PG 52-10	PG 52-16	PG 52-22	PG 52-28	PG 52-34	PG 52-40	PG 52-46
58		PG 58-16	PG 58-22	PG 58-28	PG 58-34	PG 58-40	
64	PG 64-10	PG 64-16	PG 64-22	PG 64-28	PG 64-34	PG 64-40	
70	PG 70-10	PG 70-16	PG 70-22	PG 70-28	PG 70-34	PG 70-40	
76	PG 76-10	PG 76-16	PG 76-22	PG 76-28	PG 76-34		
82	PG 82-10	PG 82-16	PG 82-22	PG 82-28	PG 82-34		

Source: AI, *Performance Graded Asphalt Binder Specification and Testing*, Superpave Series No. 1 (SP-1), The Asphalt Institute (AI), Lexington, KY, 69pp., 1994.

metre (L/m^2), is adjusted so that there is enough asphalt cement in the voids between the chips to bed the chips and seal the road, but not so much that the asphalt cement flushes to the surface and creates a dangerously slick running surface.

The use of chip seals began in New Zealand with coal tar–based seals in the 1880s [6.11]. With experience they gradually improved, but the design approach was still empirical. The work of F.M. Hanson put chip seal design on a more formal basis [6.11]. The research continued with more work in the 1950s, trials in 1965, and again with more work in the 1980s and 1990s. It culminated with *Chip Sealing in New Zealand* [6.11], published in hardcopy in 2005, and now available on the Internet [6.12]. The material in this section is drawn from that publication.

6.9.1 Early work in New Zealand

Hanson stated that a successful seal requires partial filling of voids between the chips. He found chips had voids of about 50% when laid loose. After rolling, this was reduced to about 30%, and after traffic in service, to about 20%. He determined that the asphalt cement should occupy about 65% and 70% of the final void volume. Further, he found that the depth of the chip layer was about equal to the average least dimension (ALD) of the chips. Thus, the volume of asphalt cement needed was related to the ALD.

He also found that the chips tended to break down in service but it has to be remembered that in the 1930s most heavy trucks had solid rubber tires and there still were drays with steel-rimmed wheels on the roads.

On the basis of his observations, Hanson developed a simple relation for the residual asphalt cement application rate (Equation 6.3):

$$R = \text{ALD} \times 0.2 \times 0.7 \quad \text{or} \quad R = 0.14\,\text{ALD} \tag{6.3}$$

where

R is the residual asphalt cement application rate (L/m^2)
ALD is the average least dimension of the chips (mm)

6.9.2 Refinements

Further refinements based on full-scale field trials constructed in 1965–1966 and monitored for 3 years led to the introduction of new terms to take into account the effects of traffic and the texture of the surface on which the chip seal is placed. The current form of the equation is the same, but there have been a number of modifications.

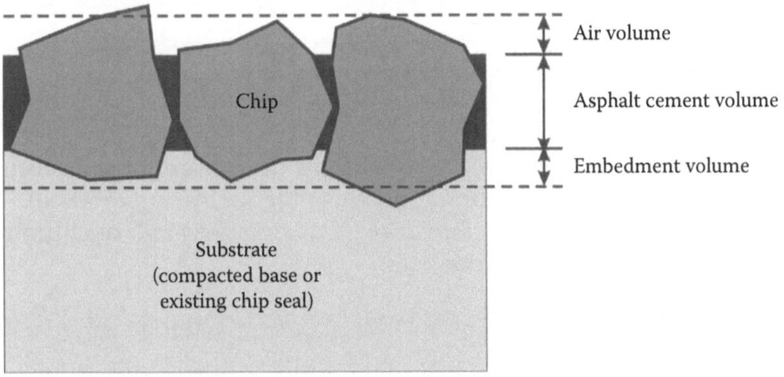

Figure 6.13 Voids concept for chip seals. (From NZTA. *Chipsealing in New Zealand*, N.Z. Transport Agency (NZTA), Wellington, New Zealand. http://www.nzta.govt.nz/resources/chipsealing-new-zealand-manual/chipsealing-in-new-zealand.html, visited September 13, 2014.)

The equation is still based on the voids concept, but an allowance is made for the partial embedment of chips into the substrate beneath (Figure 6.13). The volume of voids between the chips is related to the accumulated traffic (Equation 6.4):

$$V_v = \text{ALD}[0.83 - 0.07\log_{10}(elv)] \tag{6.4}$$

where
V_v is the volume of voids
elv is the accumulated traffic of light vehicles (passenger cars)

It is assumed that durable seals have about 60%–70% of the voids filled with asphalt cement, and that the seal achieves stability after 1 year. If the chips are not dislodged during the first cold season, there is a low risk of subsequent low temperature loss of chips. At least 35% of the voids needs to be filled with asphalt cement to avoid chip loss. A design criterion is to have a filling rate of 35% of the voids between the chips after an assumed 100 days of traffic between chip seal construction and the first cold weather period. To prevent flushing, no more than 65% of the voids should be filled with asphalt cement.

The effects of heavy truck traffic or taking into account of the truck factor T_f (Equation 6.5):

$$T_f = 1 + 0.09m \tag{6.5}$$

where
T_f is the "truck factor"
m is the percentage of the traffic volume due to heavy commercial vehicles

In addition, modifications have been made to the equation to account for the amount of asphalt cement needed to fill the texture of the substrate (new compacted granular base or existing chip seal being overlaid). The current form of the equation taking these assumptions and modifications into account is given in Equation 6.6.

$$V_b = (\text{ALD} + 0.7T_d)[0.291 - 0.025 \log_{10}(T_f \times T \times 100)] \quad (6.6)$$

where
V_b is the application rate of the asphalt cement ("binder") (L/m^2)
ALD is the average least dimension of the chips (mm)
T_d is the depth of the texture of the substrate (mm)
T_f is the truck factor (Equation 6.5)
T is the traffic (veh/lane/day)

Finally, adjustments are made to account for an absorptive substrate surface, a soft substrate, steep grades, chip shape, and the effects of urban and/or low traffic volumes. The final form of the equation as it is used in design now is given in Equation 6.7:

$$R = V_b + A_s + S_s + G_s + U_s \quad (6.7)$$

where
R is the final residual asphalt cement ("binder") application rate (L/m^2)
V_b is basic application rate (Equation 6.6) (L/m^2)
A_s is an allowance for an absorptive substrate
S_s is an allowance for a soft substrate
G_s is an allowance for steep gradients
U_s is an allowance for urban or low traffic volumes

See Chipsealing in New Zealand [6.11, pp. 341–344] for the modification factors.

6.9.3 Chip and asphalt cement selection

Determining the application rate for the asphalt cement is just one aspect of chip seal design. Consideration must be given to the selection of the asphalt cement type, the selection of the chip grade (size), the application rate for the chips, and any special circumstances such as traffic composition, when designing chip seals. See Chipsealing in New Zealand [6.11] for details.

6.10 ASPHALT CONCRETE

Asphalt concrete is a dense graded mixture of coarse and fine aggregates, a mineral filler, and asphalt cement. It is necessary to specify the mix proportions—the recipe. Developed by Bruce Marshall in the Mississippi State Highway Department in 1939, subsequently improved by the U.S. Army Corps of Engineers [6.13], and adopted around the world, the Marshall method is an empirical method for determining asphalt concrete mix designs. The Hveem method is a similar empirical method; the Marshall method [6.13] will be outlined here.

6.10.1 Basis of the Marshall method

The basic objective of the Marshall method is to select the right amount of asphalt cement to be combined with a chosen mix of coarse and fine aggregates and mineral filler. The selection of the quantity of asphalt cement is based on tests of *stability* and *flow*, and on an analysis of voids. Trial mixes are made and tested. At least three replicates of the mix at each cement content should be made, and there should be two or more cement contents above the optimum and two or more below. It is therefore typical to have about 18 mixes to test, three replicates × six asphalt cement contents. O'Flaherty [6.14] presents an equation to estimate the optimum asphalt cement content (Equation 6.8):

$$P = 0.035a + 0.045b + Kc + F \tag{6.8}$$

where

- P is the estimated optimum asphalt content (% by weight of the total mix)
- a is the percentage of the aggregate retained on the 2.36 mm sieve
- b is the percentage of the aggregate passing the 2.36 mm sieve and retained on the 75 μm sieve
- c is the percentage of the aggregate passing the 75 μm sieve
- K is 0.15 for 11%–15% passing the 75 μm sieve, 0.18 for 6%–10% passing the 75 μm sieve, and 0.20 for 5% passing the 75 μm sieve
- F is 0%–2.0%, based on the absorption of light or heavy aggregates—in the absence of other data, F = 0.7% is suggested

In the marshall method, the test results for the averages of each of the asphalt cement contents are checked against ranges of acceptable values.

6.10.2 Apparatus and procedures

Tests are performed on compacted pucks made of the asphalt concrete mixes. The compacted specimens are 102 mm (4.0 in.) in diameter and 64 mm (2.5 in.) thick, compacted using a drop hammer apparatus, usually mechanized. Depending upon the traffic for which the mix is being designed, 35, 50, or 75 blows are applied to each end of the specimen; compaction effort as a function of traffic is indicated in Table 6.10. Each compacted specimen is brought to a constant temperature of 60°C ± 1°C by immersion in a water bath for 30–40 min, then loaded across its diameter in the Marshall compression testing machine. The loading takes place at a constant rate of 51 mm/min, until failure occurs. The maximum loading, recorded in Newtons, is the Marshall stability. The deformation at that load is the flow.

Compacted specimens are rarely exactly 64 mm thick. Therefore, the stability values must be adjusted for specimen thickness or volume. Table 6.11 provides the required conversion factors.

Table 6.10 Traffic and compactive effort in Marshall method

Classification	*Traffic*	*Compactive effort: Blows to each end of specimen*
Light	ESAL < 10^4	35
Medium	10^4 < ESAL < 10^6	50
Heavy	ESAL > 10^6	75

ESAL is equivalent single axle loads (18,000 lb or 80 kN).

Table 6.11 Stability correlation values

Approximate thickness of specimen		
(mm)	*(in.)*	*Correlation coefficient*
25.4	1	5.56
31.8	1¼	3.85
38.1	1½	2.78
44.4	1¾	1.92
50.8	2	1.47
57.2	2¼	1.19
63.5	2½	1.00
68.3	2¾	0.86
76.2	3	0.76

Source: Courtesy of AI, *The Asphalt Handbook*, Manual Series No. 4 (MS-4), 1989 edn., The Asphalt Institute Inc., Lexington, KY, 607pp., 1989.

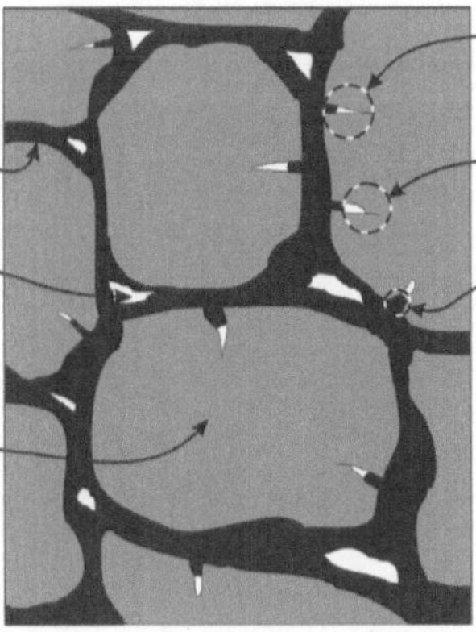

Figure 6.14 Voids in compacted asphalt concrete.

6.10.3 Voids analysis [6.13]

Beside the stability and flow, the optimum asphalt cement content is determined from an analysis of voids. It is important to understand that while there are voids between the pieces of aggregate as there is for any aggregate (recall Figure 6.1), usually there are voids within the aggregate pieces themselves, which may be partially filled with asphalt cement (Figure 6.14). As a result, to account for this absorbed asphalt cement, a slightly more complicated phase diagram is used, Figure 6.15.

The voids-related parameters used to help decide the optimal asphalt cement content are the air voids, percentage voids filled with asphalt cement, and percentage voids in the aggregate. The next sections derive the parameters needed to do the voids analysis.

6.10.3.1 Voids analysis parameters: Specific gravities

Because most aggregates are porous, they absorb water and asphalt cement. The amounts absorbed vary from aggregate to aggregate. The variation is accounted for with three definitions of specific gravity, where the definitions depend on which volumes are relevant:

- The *bulk specific gravity* (G_{sb}) is the ratio of the mass of a unit volume of the permeable aggregate, including the volumes of the permeable and impermeable voids, to the density of water (gas free, distilled, in air, at the same temperature).
- The *apparent specific gravity* (G_{sa}) is the ratio of the mass of a unit volume of the aggregate, excluding the water permeable and asphalt

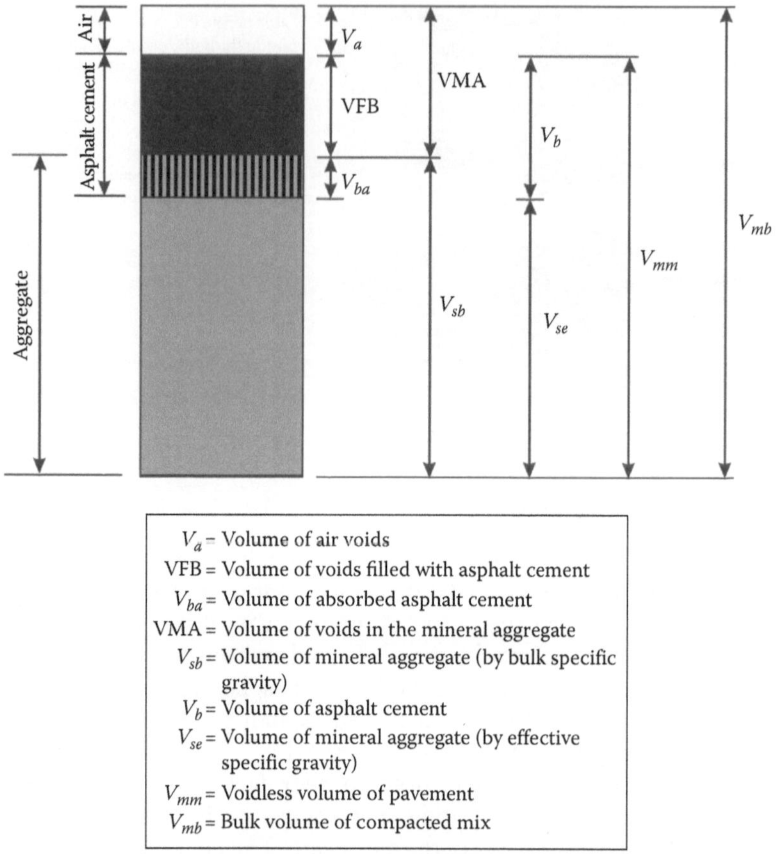

Figure 6.15 **Phase diagram for asphalt concrete.**

cement permeable volumes, to the density of water (gas free, distilled, in air, at the same temperature).

- The *effective specific gravity* (G_{se}) is the ratio of the mass of a unit volume of the aggregate, excluding the asphalt cement permeable volume, to the density of water (gas free, distilled, in air, at the same temperature).

Usually the total aggregate comprises coarse and fine aggregates and mineral filler, each with its own bulk specific gravity. In that case, the bulk specific gravity G_{sb} of the total aggregate is calculated as in Equation 6.9:

$$G_{sb} = \frac{P_1 + P_2 + \cdots + P_n}{(P_1/G_1) + (P_2/G_2) + \cdots + (P_n/G_n)} \tag{6.9}$$

where

G_{sb} is the bulk specific gravity of the total aggregate
$P_1, P_2, \ldots, P_n$ are the individual percentage masses of the coarse and fine aggregates and mineral filler
$G_1, G_2, \ldots, G_n$ are the individual bulk specific gravities of the coarse and fine aggregates and mineral filler

The bulk specific gravity of the mineral filler is often difficult to determine. If so, the apparent specific gravity of the filler can usually be substituted with little error.

The effective specific gravity G_{se} is determined from the volume of aggregate minus the void volume filled with asphalt cement. It is calculated as in Equation 6.10:

$$G_{se} = \frac{P_{mm} - P_b}{(P_{mm}/G_{mm}) - (P_b/G_b)} \tag{6.10}$$

where

G_{se} is the effective specific gravity of the total aggregate
P_{mm} is the percentage by mass of the loose mixture = 100
P_b is the percentage by mass of the asphalt cement
G_{mm} is the maximum specific gravity of the mixture (no voids)
G_b is the specific gravity of the asphalt cement

The maximum specific gravity G_{mm} is required to calculate the air voids at each asphalt cement contents. In turn, G_{mm} is dependent on the effective specific gravity, G_{se} (Equation 6.11):

$$G_{mm} = \frac{P_{mm}}{(P_s/G_{se}) + (P_b/G_b)} \tag{6.11}$$

where

G_{mm} is the maximum specific gravity of the paving mixture (no air voids)
P_{mm} is the percentage by mass of the loose mixture = 100
P_s is the aggregate content, percentage of the total mass of the paving mixture
G_{se} is the effective specific gravity of the aggregate
P_b is the percentage asphalt cement content of the paving mixture
G_b is the specific gravity of the asphalt cement

The amount of asphalt cement absorbed, P_{ba}, is calculated on the basis of aggregate mass rather than the mass of the total paving mix, Equation 6.12:

$$P_{ba} = 100\frac{G_{se} - G_{sb}}{G_{se}G_{sb}}G_b \tag{6.12}$$

where
- P_{ba} is the absorbed asphalt cement as a percentage of the mass of the aggregate
- G_{se} is the effective specific gravity of the aggregate
- G_{sb} is the bulk specific gravity of the aggregate
- G_b is the specific gravity of the asphalt cement

The effective asphalt cement content, P_{be}, is the total amount of asphalt cement in the mixture minus the amount that is absorbed into the aggregate, Equation 6.13. It is the asphalt cement that is available to coat the aggregate particles, thus governing the performance of the pavement mixture.

$$P_{be} = P_b - \frac{P_{ba}}{100}P_s \tag{6.13}$$

where
- P_{be} is the effective asphalt cement content as a percentage of the total mass of the paving mixture
- P_b is asphalt cement content as a percentage of the total mass of the paving mixture
- P_{ba} is the absorbed asphalt cement as a percentage of the mass of the aggregate
- P_s is the aggregate content as a percentage of the total mass of the paving mixture

6.10.3.2 Percent VMA in compacted paving mixture

The voids in the mineral aggregate, VMA, is the void space between aggregate particles in the compacted paving mixture including the air voids in the effective asphalt content. It is calculated on the basis of the total volume (Equation 6.14):

$$\text{VMA} = 100 - \left(\frac{G_{mb}P_s}{G_{sb}}\right) \tag{6.14}$$

where
- VMA is the voids in the mineral aggregate
- G_{mb} is the bulk specific gravity of the paving mixture

P_s is the aggregate content as a percentage of the total mass of the paving mixture
G_{sb} is the bulk specific gravity of the aggregate

6.10.3.3 *Air voids in compacted paving mixture*

The air voids of the compacted paving mixture are the small spaces between aggregate particles (Equation 6.15):

$$P_a = 100 \frac{G_{mm} - G_{mb}}{G_{mm}} \tag{6.15}$$

where
P_a is the air voids in the compacted pavement mixture, as a percentage of the total volume
G_{mm} is the maximum specific gravity of the pavement mixture
G_{mb} is the bulk specific gravity of the compacted mixture

6.10.3.4 *Percent VFA in compacted paving mixture*

The voids filled with asphalt cement is calculated (Equation 6.16):

$$\text{VFA} = \frac{100(\text{VMA} - P_a)}{\text{VMA}} \tag{6.16}$$

where
VFA is void volume filled with asphalt cement
VMA is the voids in the mineral aggregate
P_a is the air voids in the compacted pavement mixture, as a percentage of the total volume

6.10.4 Selection of optimum asphalt cement content

To select the optimum asphalt cement content, the air voids, VMA, VFA, unit weight, stability, and flow are all plotted against the asphalt cement content. The asphalt cement content for the percent air voids equal to 4% is identified. Then the graphs are consulted to find the VMA, VFA, stability, and flow values for this asphalt cement content. These are compared to the mix design criteria given in Table 6.12. If all the criteria are met, the mix is considered suitable. If one or more of the criteria are not met, adjustments must be made to the mix.

Table 6.12 Marshall mix design criteria

	Light traffic	*Medium traffic*	*Heavy traffic*
Mix design criterion	*Surface and base*	*Surface and base*	*Surface and base*
Compaction, number of blows each end of specimen	35	50	75
Stability: minimum (N | lb)	3340 | 750	5340 | 1200	8010 | 1800
Flow: maximum (0.25 mm or 0.01 in.)	18	16	14
Flow: minimum (0.25 mm or 0.01 in.)	8	8	8
Percent air voids: maximum	5	5	5
Percent air voids: minimum	3	3	3
Percent voids in aggregate, VMA	see Table 6.13		
Percent voids filled with asphalt cement, VFA: maximum	80	78	75
Percent voids filled with asphalt cement, VFA: minimum	70	65	65

Source: *Courtesy of* AI, *The Asphalt Handbook*, Manual Series No. 4 (MS-4), 7th edn., The Asphalt Institute Inc., Lexington, KY, 788pp., 2010.

Notes:

1. All criteria, not just stability value alone, must be considered in designing an asphalt paving mix. Hot mix asphalt bases that do not meet these criteria when tested at 60°C (140°F) are satisfactory if they meet the criteria when tested at 38°C (100°F) and are placed 100 mm (4 in.) or more below the surface. This recommendation applies only to regions having a range of climatic conditions similar to those prevailing throughout most of the United States. A different lower test temperature may be considered in regions having more extreme climatic conditions.
2. Traffic classifications: Light—traffic conditions resulting in a design ESAL < 10^4. Medium—traffic conditions resulting in a design ESAL between 10^4 and 10^6. Heavy—traffic conditions resulting in a design ESAL > 10^6.
3. Laboratory compaction efforts should closely approach the maximum density obtained in the pavement under traffic.
4. The flow value refers to the point where the load begins to decrease.
5. The portion of asphalt cement lost by absorption into the aggregate particles must be allowed for when calculating percent air voids.
6. Percent voids in the mineral aggregate should be calculated on the basis of the ASTM bulk specific gravity for the aggregate.

6.10.5 Superpave

In the 1990s, the United States and Canada adopted a new, performance-based approach to asphalt mix design called Superpave (SUPerior PERforming asphalt PAVEments [6.15]). The Superpave approach includes new ways of specifying asphalt cements and mineral aggregates; developing asphalt mixture designs; and analyzing and predicting pavement performance and comprises new methods, equipment, and software. Because of its complexity and expense compared to the Marshall method, it is not covered here.

Table 6.13 Minimum value of voids in mineral aggregate (VMA)

Nominal maximum particle size		*Design air voids, %*		
(mm)	*(in.)*	*3.0*	*4.0*	*5.0*
1.18	0.046	21.5	22.5	23.5
2.36	0.093	19.0	20.0	21.0
4.75	0.19	16.0	17.0	18.0
9.50	0.37	14.0	15.0	16.0
12.5	0.50	13.0	14.0	15.0
19.0	0.75	12.0	13.0	14.0
25.0	1.00	11.0	12.0	13.0
37.5	1.50	10.0	11.0	12.0

Source: O'Flaherty, C.A., *Highways—The Location, Design, Construction & Maintenance of Pavements*, Butterworth-Heinmann, Oxford, U.K., 553pp., 2002.

6.11 ROLLER-COMPACTED CONCRETE

Roller-compacted concrete (RCC) is stiff, zero-slope concrete placed and compacted using conventional asphalt paving equipment [6.16]. It is constructed with no formwork and has no reinforcement, joints, or dowels. RCC was first developed for dams and paving and its first application for paving was the construction of a log sorting yard on Vancouver Island in 1976 [6.16]. When used in industrial applications such as roads, laydown areas, dockside pavements, and log sorting yards, RCC pavements have the advantages of resisting rutting, not deteriorating after spills of fuel or hydraulic fluids, and not softening in hot weather [6.17]. They are comparatively inexpensive to construct because they use conventional asphalt paving equipment and methods.

RCC mixes should be designed to provide maximum density. There should be just enough cement paste to fill the voids in the aggregates. It is important to mix RCC thoroughly to disperse the cement paste, as there is relatively little of it [6.16]. Mixing is best achieved with continuous-flow pug mills. Compared to regular Portland cement concrete, RCC has a lower moisture content, lower cement paste content, and requires a lower fine aggregate content to be stable under vibratory rolling [6.16]. The maximum aggregate size should be 20 mm (¾ in.) to provide a good riding surface but lower layers in multilift pavements, not exposed to traffic, can have a maximum aggregate size of 40 mm (1⅝ in.) [6.16].

Thickness design is achieved with methods similar to those for conventional Portland cement concrete pavements. The methods depend upon such variables as modulus of subgrade reaction, design axle loads, and the flexural strength and modulus of elasticity of the RCC [6.16].

As with conventional concrete pavements, RCC requires a good base and subbase. Once these are prepared, the concrete is trucked to site in covered dump trucks and transferred to conventional- or high-density asphalt pavers. RCC should be placed in layers no less than 100 mm (4 in.) thick nor more than 200 mm (8 in.) thick [6.16]. Compaction is very important, as it determines the strength, durability, and water shedding capabilities of the RCC. Compacted near the optimum moisture content to the maximum density, RCC should be dry enough to support vibratory rollers, yet wet enough to allow the complete dispersion of the cement paste. It is not necessary to saw cut joints unless it is thought needed for appearance's sake. As with conventional concrete, curing of the RCC is important, to allow the proper hydration of the cement. It is typically cured for at least seven days, by spraying with water, covering with wet burlap, or by spraying a polymer membrane. Apart from the necessary water truck, RCC should not be opened to traffic until the required flexural strength has been reached.

REFERENCES

6.1 Lambe, T.W. and Witman, R.V. 1979. *Soil Mechanics, S.I. Version*. Toronto, Ontario, Canada: John Wiley & Sons, 553pp.

6.2 Wright, P.H. and Paquette, R.J. 1987. *Highway Engineering*, 5th edn. Toronto, Ontario, Canada: John Wiley & Sons, 717pp.

6.3 MTO. 1999. *Laboratory Testing Manual*. Toronto, Ontario, Canada: Ministry of Transportation of Ontario (MTO). *See also* http://applications.roadauthority.com/Standards/?Id=4EF3A115–7A19–4BC1–9BBF-689A6C8EC222. August 23, 2014.

6.4 MTO. Ontario provincial standards for roads & public works: OPSS 1010 material specification for aggregates-base, subbase, select subgrade, and backfill material. Toronto, Ontario, Canada: Ministry of Transportation of Ontario. http://www.raqsb.mto.gov.on.ca/techpubs/ ops.nsf/d37f5a16d8174ffa85256d130066857f/702bdff1eaaf4d9e85257b5c006080ac?OpenDocument, visited August 23, 2014.

6.5 Alberta Infrastructure and Transportation. 2007. Standard specifications for highway construction. Edmonton, Alberta, Canada: Professional Services Section, Alberta Infrastructure and Transportation, p. 230. http://www.transportation.alberta.ca/Content/docType245/Production/Specs2007.pdf, visited September 2, 2014.

6.6 British Columbia Ministry of Transportation and Infrastructure. 2012. Standard specifications for highway construction. Victoria, British Columbia, Canada: BC Ministry of Transportation and Infrastructure. http://www.th.gov.bc.ca/publications/const_maint/contract_serv/standard_specs/SS202_Table_202-C-Aggregate_Gradations.pdf, visited September 2, 2014.

6.7 Ferry, A.G. 1986. *Unsealed Roads, a Manual of Repair and Maintenance for Pavements*. Wellington, New Zealand: Road Research Unit, New Zealand National Roads Board.

6.8 Rodriguez, A.R., del Castillo, H., and Sowers, G.F. 1988. *Soil Mechanics in Highway Engineering*. Clausthal-Zellerfeld, Germany: Trans Tech Publications, 843pp.

6.9 AI. 2010. *The Asphalt Handbook*. Manual Series No. 4 (MS-4), 7th edn. Lexington, KY: The Asphalt Institute, 788pp.
6.10 AI. 1994. *Performance Graded Asphalt Binder Specification and Testing*. Superpave Series No. 1 (SP-1). Lexington, KY: The Asphalt Institute (AI), 69pp.
6.11 Transit New Zealand. 2005. *Chipsealing in New Zealand*. Wellington, New Zealand: Transit New Zealand (now N.Z. Transport Agency), 524pp.
6.12 NZTA. 2014. *Chipsealing in New Zealand*. Wellington, New Zealand: N.Z. Transport Agency (NZTA). http://www.nzta.govt.nz/resources/ chipsealing-new-zealand-manual/chipsealing-in-new-zealand.html, visited September 13, 2014.
6.13 AI. 1989. *The Asphalt Handbook*. Manual Series No. 4 (MS-4), 1989th edn. Lexington, KY: The Asphalt Institute, 607pp.
6.14 O'Flaherty, C.A. 2002. *Highways—The Location, Design, Construction & Maintenance of Pavements*. Oxford, U.K.: Butterworth-Heinmann, 553pp.
6.15 AI. 1995. *Superpave Level 1 Mix Design*. Superpave Series No. 2 (SP-2). Lexington, KY: The Asphalt Institute, 607pp.
6.16 CPCA. undated. Roller compacted concrete (RCC) pavements design and construction. CPCA CP018.01P. Ottawa, Canada: Canadian Portland Cement Association (CPCA). http://www.cement.ca/images/stories/CP018.01P.pdf, visited October 26, 2014.
6.17 CAC. RCC pavement roller-compacted concrete. Ottawa, Ontario, Canada: Cement Association of Canada (CAC). http://www.cement.ca/en/Highways/RCC-Pavement-Roller-compacted-concrete.html, visited October 26, 2014.
6.18 Bowles, J.E. 1992. *Engineering Properties of Soils and Their Measurement*, 4th edn. Toronto, Ontario, Canada: McGraw-Hill Inc. 241 pp. + data sheets.
6.19 Lambe, T.W. 1951. *Soil Testing for Engineers*. New York: John Wiley & Sons, Inc., 165pp.
6.20 Wray, W.K. 1986. *Measuring Engineering Properties of Soils*. Englewood Cliffs, NJ: Prentice-Hall Inc., 276pp.

Chapter 7

Pavement design

7.1 INTRODUCTION

Pavement design is primarily a matter of selecting the thicknesses and materials of the layers that comprise the pavement. The designer selects the materials for the pavement layers, usually placing the strongest materials toward the top of the pavement where the stresses imposed by the wheel loads are greater. A top-down approach is taken, with each layer protecting the weaker layer below.

Beside purely mechanical considerations, other issues such as constructability, economics, sustainability, and the environment come into play in pavement design.

The discussion here will be limited to flexible pavements including asphalt, surface-treated, and aggregate-surfaced pavements.

7.2 INPUTS TO PAVEMENT DESIGN

The mechanical considerations for pavement design require a number of key inputs, including the traffic loading, the strength of the subgrade (see Chapter 5 for definitions of pavement components), strengths of the pavement layer materials, climatic effects, and the failure criterion for the design.

7.2.1 Traffic

The traffic is specified in terms of the numbers of axle passes and of the axle loading. To cope with the wide spectrum of vehicle configurations (numbers, placement, and spacing of axles) and axle loads, which can range from that of a light passenger vehicle to as much as 400 tonnes (440 tons) for heavy mining trucks, the concept of *equivalent single axle loads* (ESAL) is introduced. In the ESAL concept, the increment of pavement life used up by one pass of the axle is equated to the increment of pavement life used up by a number of passes of a "standard axle." The standard axle load is

80 kN (18,000 lb). The pavement life used by the passage of axles is far from linear with the number of passes, rather, it is generally taken to be related to the fourth power of the axle load ratio:

$$\text{ESAL} \cong \left(\frac{W}{W_s}\right)^4 \tag{7.1}$$

where

ESAL is the number of equivalent single axle loads per pass of the axle
W is the axle load (kN or lb)
W_s is the standard axle loading (80 kN if working in SI units, 18,000 lb if working in U.S. Customary Units)

This is a simplistic approach to the determination of the design traffic volume, as there are many factors that affect the increment of damage caused by the passage of an axle or group of axles. The damage is related to the fatiguing stress reversal in the pavement layers with the approach and departure of each axle, so the individual axle loads, number of axles, and the spacing of the axles in a group are important factors. The index may be other than 4 for various pavement types, especially gravel-surfaced pavements. In addition, there is an interaction between the condition of the pavement and the damaging effect of the passage of an axle group. AASHTO [7.1] provides tables for ESAL as a function of numbers of axles in a group (single axle, tandem axles, triple axles) and the terminal present serviceability index,* PSI, a measure of pavement roughness.

Beyond these issues, just estimating the number of vehicles to use the road during its service life can be a challenge. There are a number of components, each often difficult to assess, in the traffic estimate, including the following:

- *Existing traffic*: The current number of vehicles using the road can vary from hour to hour during the day, day to day during the week, and season to season during the year. To take into account these variations for pavement design, the traffic volume is usually expressed as the annual average daily traffic, AADT. The AADT is the sum of vehicles passing a particular point on the road in both direction over 365 consecutive days, divided by 365.
- *Traffic growth*: As the use of the land surrounding the road, the population, and the wealth of the population all increase, the traffic volume can be expected to increase generally each year. A growth

* The U.S. "present serviceability index," *PSI*, is a subjective rating of pavement roughness, ranging from 0 ("very poor") to 5 ("very good"). The Canadian equivalent is "riding comfort index," *RCI*, ranging from 0 to 10. See Haas et al. [7.2].

factor is usually applied to the existing traffic volume, and the increase compounded annually over the service life of the road. Usually growth is observed, but a decline in volume is also conceivable.

- *Development growth*: The traffic on the road may increase due to some specific nearby development that takes place during the life of the road. For instance, a new agricultural processing plant may be constructed nearby, or a housing subdivision (estate) built, or a new quarry opened up, any of which will result in an increase in traffic using the road.
- *Attracted traffic*: Improving a road may make it a more attractive route than some other route currently being used, and if so, some of the traffic will divert to the new or improved road. On the other hand, conceivably some other route may become more attractive, resulting in diversion of traffic *away* from the road.
- *Generated traffic*: Improving a road may encourage vehicle operators to use it, where before they never used the road to ship goods or travel, or perhaps used some other mode (e.g., foot or bicycle paths, rail, etc.).

Taking these factors into account can be a challenge when carrying out pavement designs for public roads. However, estimating traffic for resource access roads, such as haul roads for forest or mining operations, can be easier. Usually the life of the facility, the amount of the resource (e.g., volume of timber, tonnage of ore) to be moved, and the capacity of the haul trucks are all known, making the estimation of the number of passes of the design vehicle a relatively simple matter.

7.2.2 Subgrade strength

The reason to build a pavement is to protect the weak subgrade; the weaker the subgrade, the more substantial the pavement will have to be. Subgrade strength can be measured with tests on laboratory specimens, or *in situ* (in place at the site). There are advantages and disadvantages to both. Laboratory testing permits the designer to control the parameters that affect soil behavior, but there is a risk of sample disturbance which will affect the results. *In situ* testing may give more direct estimates of the subgrade strength, but the cost of testing can be high, and access to the site is required.

Many of the common pavement design methods are based on the California bearing ratio (CBR). The CBR test compares the penetration resistance of a specimen of the subgrade soil to that of high quality compacted crushed rock [7.3]. In the test (Figure 7.1), a hardened steel cylinder 49.63 mm (1.95 in.) in diameter is pressed into a prepared specimen of the subgrade soil 152 mm (6.0 in.) in diameter by 178 mm (7.0 in.) height at a rate of 1.3 mm/min (0.05 in./min). The "CBR" of the specimen is the ratio

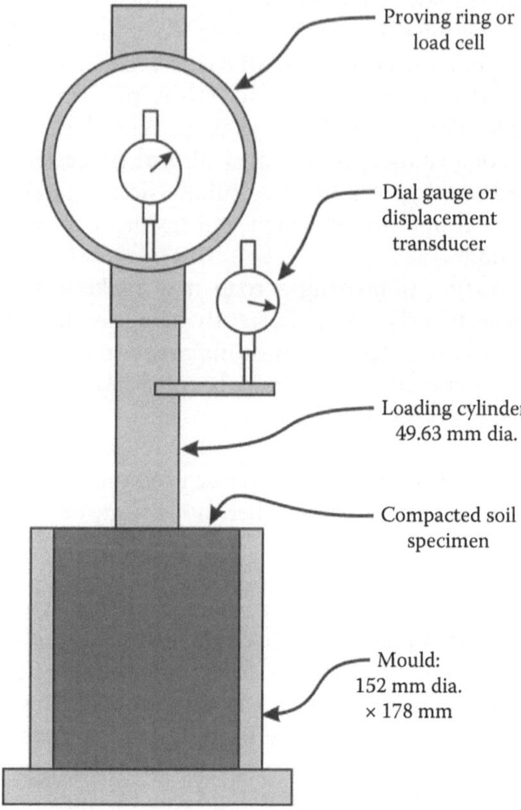

Figure 7.1 Schematic diagram of the CBR laboratory test apparatus. Proving ring or load cell measures the force applied to the loading cylinder, dial gauge or displacement transducer measures its penetration into the soil. (Not to scale.)

in percent of the pressure exerted on the specimen divided by a standard pressure at 2.5 mm (0.1 in.) and 5.0 mm (0.2 in.) penetration. The standard pressures are 6.9 MPa (1000 lb/in.^{2}) and 10.3 MPa (1500 lb/in.^{2}) for 2.5 and 5.0 mm penetration, respectively. The value of the CBR is usually the ratio calculated for 2.5 mm penetration. If the value is greater for 5.0 mm penetration, the test is run again, and if the value at 5.0 mm penetration is still greater than that at 2.5 mm penetration, the value at 5.0 mm penetration is taken as the final value of the CBR.

Care must be taken to obtain representative samples of the subgrade soil. The soil should be compacted to the same density, at the same moisture content, as the subgrade soil will have during the service life of the pavement.

Laboratory CBR tests are time consuming and expensive. The test with the same loading apparatus can be conducted in the field, but the cost is high for the little data obtained. For these reasons, more often designers carry out

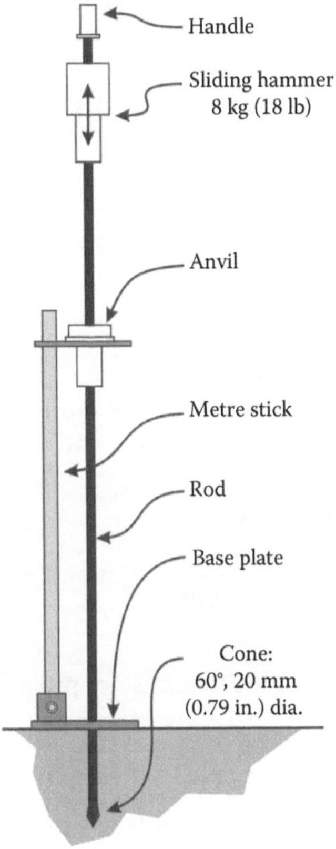

Figure 7.2 Schematic diagram of the manual DCP test apparatus. Sliding 8 kg (18 lb) hammer travels 575 mm (22.6 in.), striking the anvil. Penetration is measured by the anvil bracket's progress down the metre stick. (Not to scale.)

many replicates of some less expensive, rapid field test which is correlated to CBR. The test frequently chosen is the manual dynamic cone penetration (DCP) test (Figure 7.2). In the DCP test, a 20 mm (0.80 in.) diameter 60° cone is driven into the subgrade by an 8 kg (18 lb) hammer falling 575 mm (22.6 in.) [7.4]. The blow count in mm/blow has been correlated to CBR [7.4]. A graph of blow count versus depth is recorded for each test, and an appropriate subgrade CBR determined from the correlation. The test is rapid and relatively inexpensive to perform, but the manual version of the apparatus requires 2–3 operators. Where labor is inexpensive, the labor intensiveness of the test is not a problem; where labor *is* expensive, a mechanized version of the apparatus requiring just one operator can be used.

Other apparatus such as the Clegg hammer or light weight deflectometer can be used. The outputs of these are correlated to CBR and soil modulus, respectively.

In many pavement designs, subgrade resilient modulus M_r is the measure of the subgrade strength used. M_r is the slope of the stress–strain curve in a repeated-load triaxial test of the subgrade [7.5]. A design value of subgrade M_r can be correlated* to CBR. Common correlations are [7.1]:

$$M_r(\text{psi}) = 1500 \text{ CBR} \tag{7.2}$$

or

$$M_r(\text{MPa}) = 10 \text{ CBR} \tag{7.3}$$

Recognizing that there is a limit to the relation, NAASRA [7.6,7.7] proposed curvilinear relationships between M_r and CBR:

$$\text{For CBR} < 5: M_r = 16.2 \text{ CBR}^{0.7} \tag{7.4}$$

$$\text{For } 5 < \text{CBR} < 15: M_r = 22.4 \text{ CBR}^{0.5} \tag{7.5}$$

Preliminary values for subgrade strengths are provided in Table 7.1 [7.8].

7.2.3 Pavement layer material strengths

Values for the strengths of the pavement layers are either assumed or determined for pavement design. For finite element-based design methods, elastic moduli are usually used. The AASHTO design method defines a structural layer coefficient, a_i for each layer (see Table 7.2).

7.2.4 Climatic effects

Pavement design methods must take into account the effects of climate, particularly temperature and precipitation. Subgrade strength decreases markedly in wet periods and a large proportion of the pavement life can be used up when traffic is even a short, wet period or during the thaw in the spring of the year in temperate climates. Frozen pavements are obviously much stronger than those that are not and deteriorate little with traffic during periods of freezing temperatures. The more sophisticated pavement design methods take into account the state of the climate when traffic is applied.

* Which in turn is often correlated to DCP penetration/blow—thus resulting in design which relies on a correlation to a correlation!

Table 7.1 Recommended M_r values for subgrades

Description	MTO classification	Drainage characteristics	Susceptibility to frost action	Resilient modulus M_r for subgrade conditions (MPa)			Resilient modulus M_r for subgrade conditions (lb/in.^{2})		
				Good	Fair	Poor	Good	Fair	Poor
Rock, rock fill, shattered rock, boulders/cobbles	Boulders/cobbles	Excellent	None	90	80	70	13,100	11,600	10,200
Well-graded gravels and sands suitable as granular borrow	GW, SW	Excellent	Negligible	80	70	50	11,600	10,200	7,300
Poorly graded gravels and sands	GP, SP	Excellent to fair	Negligible to slight	70	50	35	10,200	7,300	5,100
Silty gravels and sands	GM, SM	Fair to semi-impervious	Slight to moderate	50	35	30	7,300	5,100	4,400
Clayey gravels and sands	GC, SC	Practically impervious	Negligible to slight	40	30	25	5,800	4,400	3,600
Silts and sandy silts	ML, MI	Typically poor	Severe	30	25	18	4,400	3,600	2,600
Low plasticity clays and compressible silts	CL, MH	Practically impervious	Slight to severe	35	20	15	5,100	2,900	2,200
Medium to high plasticity clays	CI, CH	Semi-impervious to impervious	Negligible to severe	30	20	15	4,400	2,900	2,200

Source: Reproduced with permission from Hajek, J.J., Smith, K.L., Rao, S.P., and Darter, M.I., Adaptation and verification of AASHTO pavement design guide for Ontario conditions, Ministry of Transportation of Ontario, Pavements and Foundation Section, Prepared by ERES Consultants, Applied Research Associates, Inc., Toronto, Ontario, Canada, 2008 © Queen's Printer for Ontario, 2015.

If a pavement design method requires that only a single value of subgrade strength be declared, the value must be very carefully selected, to represent the most damaging season in the design life of the pavement during which there is traffic.

7.2.5 Failure criterion and reliability

Pavement design methods must have an expressed or inherent definition of "failure." This can be a maximum rut depth, or a maximum strain, or the terminal state of the pavement (e.g., measured by pavement roughness).

The more sophisticated pavement design methods also recognize the notion of design reliability. Because of the variability of subgrade strength and pavement material strengths along the road, there will be places where the design is inadequate. A design reliability in percent reflects the designer's tolerance to pavement failure. In the context of low-volume roads, lower reliabilities than those expected for high-volume highways are tolerated. The higher the reliability, the thicker the pavement. A reliability of 100% implies that the road is overbuilt for most of its length, which is economically unacceptable.

7.3 OTHER CONSIDERATIONS

Beside the purely mechanical, other considerations need to be addressed in pavement design.

The most economical design is desired but that may not necessarily imply that the least expensive pavement be constructed. There is an interaction between the design of the pavement and its maintenance. A pavement with a low construction cost could ultimately lead to expensive maintenance and rehabilitation, so the *whole life* cost of the pavement must be taken into account. In addition, there is an interaction between pavement design and vehicle operating costs. A pavement constructed at minimal cost may result in elevated vehicle operating costs, including fuel consumption and vehicle maintenance.

Pavements must be constructable. They must not require expensive materials, unavailable construction equipment, or uncommon construction methods. They must not be overly complicated to construct, or logistical problems will result.

There is a growing awareness of sustainability. Pavements are constructed of nonrenewable materials, and the supply of aggregates is not unlimited. Designs which permit the recycling of materials during the life of the pavement are becoming more common. Designers are also beginning to accept the use of marginal but still adequate materials, rather than specifying the best material in every case even when not warranted.

Designers are becoming more conscious of the impact of road and pavement construction on the environment, and more environmentally friendly solutions such as warm asphalt, which requires less energy to produce and results in the production of less pollution, are being implemented more often.

7.4 TYPES OF PAVEMENT DESIGN METHODS

Pavement designs fall into a handful of classes, depending upon their sophistication.

The "design" may simply be based on precedent: "this works, that does not—we'll do it this way." This is the simplest approach and is often implemented. Many road authorities have a standard suite of pavement designs based on road classification. Once the designs are calibrated, the approach works well, unless and until new conditions are encountered.

Pavement designs may be presented in simple design catalogs—tables of pavement layer design thicknesses which are functions of traffic classes and subgrade strength classes. Traffic is expressed in ESAL, and subgrade strength may be expressed in CBR classes. Inherent in this approach are assumptions about material strengths and the failure criterion.

Various CBR charts exist. These are graphs of pavement thickness plotted against either traffic in ESAL with the subgrade CBR as a parameter or plotted against subgrade CBR with traffic in ESAL as a parameter. It is assumed that the pavement has a single layer of high-quality crushed rock. Other materials can be accommodated with the concept of equivalent base thickness, where the thickness of a given layer is adjusted according to its strength compared to that of crushed rock base material.

Mechanistic-empirical pavement design methods combine some consideration of the mechanics of pavement behavior with empirical data or experience. Various design equations and charts present the results.

Selected examples of the types of pavement design methods are given below.

7.5 SELECTED PAVEMENT DESIGN METHODS

7.5.1 AASHTO methods

Low-volume road pavement design methods can be found in the *1993 AASHTO Guide for Design of Pavement Structures* [7.1]. Their origins can be traced back to the venerable AASHO Road Test [7.1,7.9,7.10], where, starting in 1958 at a site west of Chicago, near Ottawa, Illinois, 126 Army trucks of varying configuration trafficked six pavement test loops with 836 individual pavement designs for 19 h daily for 2 years. The methods include design catalogs and empirical methods.

Table 7.2 Recommended layer coefficient, a_i values for pavement materials

Pavement material	*Layer coefficient, a_i*
Bituminous bound materials	
New and recycled hot mix asphalt	0.42
Existing hot mix asphalt	0.14–0.28
Cold mix asphalt	0.28–0.38
Recycled asphalt pavement (RAP) Granular A blend stabilized with expanded asphalt cement	0.20–0.25
Existing cold mix asphalt	0.11–0.24
Bituminous-treated Granular A (about 3% asphalt cement)	0.31
Unbound materials	
New Granular A	0.14
Existing Granular A	0.10–0.14
New Granular B Type I	0.09
Existing Granular B Type I	0.05–0.09
New Granular B Type II	0.09–0.14
Existing Granular B Type II	0.06–0.14

Source: Reproduced with permission from Hajek, J.J., Smith, K.L., Rao, S.P., and Darter, M.I., Adaptation and verification of AASHTO pavement design guide for Ontario conditions, Ministry of Transportation of Ontario, Pavements and Foundation Section, Prepared by ERES Consultants, Applied Research Associates, Inc., Toronto, Ontario, Canada, 2008 © Queen's Printer for Ontario, 2015.

Note: Granular A, B, and O are MTO material designations: see Tables 6.4 and 6.5, and MTO [7.12].

For single layer aggregate-surfaced roads, the Guide [7.1] presents a catalog of layer thicknesses in inches. These are a function of the traffic levels shown in Table 7.3, five levels of the relative quality of the subgrade soil, and the location of the road's site in one of six U.S. climatic regions. Note that it is assumed the resilient modulus of the aggregate pavement layer is 30,000 lb/in.^{2}. The failure criterion is an overall design serviceability loss ΔPSI of 3.0.

Table 7.3 Traffic levels for AASHTO low-volume aggregate-surfaced road pavement design catalog

Designation	*Traffic volume (ESAL)*
High	60,000–100,000
Medium	30,000–60,000
Low	10,000–30,000

Source: AASHTO, *AASHTO Guide or Design of Pavement Structures*, American Association of State Highway and Transportation Officials (AASHTO), Washington, DC, 1993.

In a second, more general method accommodating multiple pavement layers, the Guide [7.1] defines a structural number, *SN*, a measure of overall pavement strength:

$$SN = a_1 D_1 + a_2 D_2 + a_3 D_3 \tag{7.6}$$

where

a_1, a_2, and a_3 are layer coefficients (Table 7.2) for the surface, base, and subbase layers, respectively, reflecting their material strengths; and

D_1, D_2, and D_3 are the thicknesses of the surface, base, and subbase layers, respectively.

Catalogs of *SN* values are provided, one each for reliabilities of 50% and 75%. These are also a function of traffic level (Table 7.4), relative quality of subgrade soil, and location within one of six U.S. climatic zones. The failure criterion is a terminal present serviceability index PSI_t of 1.5. With the *SN* value in hand, and known values for the layer coefficients a_i, designers task is to adjust the pavement thicknesses D_i to satisfy Equation 7.6.

A third, nomograph-based design method for single-layer pavements is presented in the Guide [7.1]. The nomograph takes into account the following parameters:

- Allowable rut depth
- Subgrade resilient modulus
- Resilient modulus of the aggregate layer
- Design traffic volume in ESAL

The amount of damage in each of four seasons (winter—frozen, spring thaw—saturated, spring/fall (autumn)—wet, and summer—dry) is calculated and checked against two criteria: rutting, and loss of serviceability.

Table 7.4 Traffic levels for AASHTO flexible pavement design catalog

Designation	*Traffic volume (ESAL)*
High	700,000–1,000,000
Medium	400,000–600,000
Low	50,000–300,000

Source: AASHTO, *AASHTO Guide or Design of Pavement Structures*, American Association of State Highway and Transportation Officials (AASHTO), Washington, DC, 1993.

The more general flexible pavement design method the Guide [7.1] presents can also be used for low-volume road pavement design. This is also based on a nomograph which takes into account the following parameters:

- Reliability
- Overall standard deviation of the data
- Estimated total ESAL over the design life of the pavement
- Subgrade resilient modulus
- Design serviceability loss

The output of the nomograph is the structural number, *SN*, which in turn is used in Equation 7.6. The nomograph is based on the following Equation [7.1]:

$$\log_{10}W_{18} = Z_r s_o + 9.36 \log_{10}(SN+1) - 0.20 + \frac{\log_{10}[\Delta PSI/(4.2-1.5)]}{0.40 + \left(1094/(SN+1)^{5.19}\right)} + 2.32\log_{10}M_r - 8.07 \quad (7.7)$$

where

W_{18} is the estimated traffic (ESAL)
Z_r is standard normal deviate (determines reliability—see p. I-63 of the Guide [7.1])
s_o is the overall standard deviation of the input data
SN is the structural number (inches)
ΔPSI is the loss in serviceability over the life of the road
M_r is the resilient modulus of the subgrade (lb/in.^{2})

Note that the equation is valid only for U.S. Customary Units

The equation can be readily programmed into a spreadsheet. Because the equation is written with the design traffic as a function of the structural number *SN*, whereas usually the traffic is known and the structural number is to be determined, an iterative approach can be adopted. A trial structural number is assumed, and Equation 7.7 solved to calculate the design traffic volume. Based on a comparison of the calculated traffic volume and the known design traffic volume, a new structural number is input to Equation 7.7, a new traffic volume calculated, and the cycle continued until convergence on the design structural number, *SN*. Design then proceeds to Equation 7.6 and the pavement layer thicknesses determined.

The Ministry of Transportation of Ontario (MTO) has published pavement design catalogs for "King's Highways and Freeways" and "Secondary Highways" in the so-called Blue Book [7.11]. Ontario-specific values for input to the AASHTO design equation (Equation 7.7) are given in Hajek et al. [7.8], including a_i values (Table 7.2).

7.5.2 British pavement design for tropical and subtropical countries

A pavement design catalog for bitumen-surfaced roads is available in *Overseas Road Note 31* [7.4]. Table 7.5 lists the types of pavements included in the design catalog. The catalog covers traffic volumes from 300,000 to 30 million ESAL.

7.5.3 Australian design chart

Australian and New Zealand practice uses either a mechanistic-empirical design approach or a design chart. The design chart is appropriate for low-volume roads consisting of a high quality compacted crushed rock base topped with a thin, nonstructural bituminous surface of surface treatment or asphalt 50 mm or less in thickness. An early version of the chart appears in Austroads [7.13]. It has the thickness of the granular base plotted against the design traffic, ranging from 10^5 ESALs to 10^8 ESALs. Curves are shown for subgrade CBR values ranging from 2 to >30, and the minimum base thickness is 100 mm. No indication of the strength of the base material or the failure criterion is given.

A similar chart is presented in ARRB [7.14], an unsealed roads manual which includes a New Zealand supplement. The chart is similar in form to that in Austroads [7.13], with base layer thickness plotted against traffic for values of subgrade CBR from 3 to >30, but the range of traffic is lower, running from 10^3 to 5×10^5 ESAL.

More recently, the same chart appears in Hoque and Jameson [7.15]. The thickness of the granular base is given on the chart as Equation 7.8:

$$t = 0.475 \times \left[219 - 211 \log \mathrm{CBR} + 58(\log \mathrm{CBR})^2\right] \times \log 14 V_t \qquad (7.8)$$

where

t is the granular base layer thickness, with a minimum of 100 mm (mm)
CBR is the subgrade CBR
V_t is the traffic volume (ESAL)

The equation is only valid for SI units. Note that the equation is incorrect as published in the reference.

7.5.4 Comparison of methods

While it is tempting to compare the results of different pavement design methods for the same inputs, the comparisons may not always be fruitful for the following reasons:

- The methods may be applicable to specific sites or conditions

Table 7.5 Pavement designs in overseas road note 31 pavement design catalog

Surface	*Base*	*Pavement designs available for traffic level*							
Surface treated	Granular	T1	T2	T3	T4	T5	T6		
Surface treated	Composite (unbound and cemented)	T1	T2	T3	T4	T5	T6	T7	
Semistructural	Granular			T3	T4	T5	T6		
Semistructural	Composite (unbound and cemented)			T3	T4	T5	T6	T7	T8
Structural	Granular						T6	T7	T8
Structural	Composite (unbound and cemented)						T6	T7	T8
Semistructural	Bituminous				T4	T5	T6	T7	T8
Surface treated	Cemented	T1	T2	T3	T4	T5	T6		

Traffic levels	**Million ESAL**
T1	<0.3
T2	0.3–0.7
T3	0.7–1.5
T4	1.5–3.0
T5	3.0–6.0
T6	6.0–10
T7	10–17
T8	17–30

Source: TRL, A guide to the structural design of bitumen-surfaced roads in tropical and subtropical countries, *Overseas Road Note 31*, 4th edn., Transportation Research Laboratory (TRL), Crowthorne, U.K., 75pp. http://www.transport-links.org/transport_links/filearea/publications/1_716_Microsoft%20Word%20-%20Overseas%20Road%20Note%2031%20edit1.pdf, visited January 15, 2015, 1993.

- The range of validity of the inputs to the design methods may be different
- There may be different inherent assumptions about input parameters
- Assumptions may not be explicit
- A parameter included in one method may not be included in another
- Failure criteria may be different

For these reasons, it should not be surprising that using the same inputs with two different pavement design methods does not necessarily yield the same resultants.

Designers are advised to adopt a method which suits their conditions and calibrate that method against existing pavements that have been successful and others that have failed in service.

7.6 U.S. MECHANISTIC-EMPIRICAL DESIGN METHOD

There has recently been a desire to develop a pavement design method which is based more on the mechanisms of pavement behavior, rather than on empiricism. Efforts have been made over the past decade in the United States to develop a mechanistic-empirical design method, culminating in the *AASHTO Mechanistic-Empirical Pavement Design Guide* (MEPDG) and associated software [7.16,7.17].

Moving from the previous empirical method described in Section 7.5.1 to the MEPDG and associated software presents the following advantages [7.17]:

- The evaluation of a broader range of vehicle loadings, material properties, and climatic effects
- Improved characterization of the existing pavement layers
- Improved reliability of pavement performance predictions

However, there is an increase in the quantity and complexity of the input data, and it is expected that implementing the MEPDG will increase the time required for pavement design and evaluation.

The mechanistic-empirical design method can be carried out at one of three levels [7.17]:

- *Level 1*: inputs are measured values, including site-specific traffic. This level requires the greatest parameter knowledge, and the greatest time, resources, and cost to obtain.
- *Level 2*: inputs are calculated from correlations with other, more easily obtained data.
- *Level 3*: inputs are based on default values, regional values, or expert opinion.

The software is arranged in modules which the designer uses to analyze a trial pavement design. The pavement structure, performance criteria, input data such as traffic, climate, layer properties, and calibration factors are input for the trial design, and the modeled behavior checked against the performance criteria. If all are acceptable, the pavement design is tentatively accepted, subject to an analysis of other factors such as whole-life costs and constructability.

The MEPDG approach, while being much more realistic than the AASHTO empirical approach presented in the 1993 Guide [7.1] seems unduly complicated and onerous for application at this time to LVR. However, it is still in development and bears monitoring over the next few years.

7.7 HEAVY-DUTY PAVEMENT DESIGN FOR HAUL ROADS

The designs presented in the previous sections have all applied to pavements designed for conventional axle loads, circa 8–10 tonnes (18,000–22,000 lb). But what about pavements designed for the very heavy hauling in forestry, mining, oil sands, or oil and gas operations? The axle loads for the vehicles on these roads can range up to hundreds of tons, "running off the chart" for conventional pavement design methods. Designers of these roads must look to other, completely different pavement design methods. It takes a completely different approach; the pavements can be extremely thick.

7.7.1 "Pavements" versus "Embankments"

When does a pavement become so thick it starts to look like an embankment? It depends on the circumstances.

Pavements are typically thin structures, where the live loading of the vehicles is the primary concern. Embankments are typically thick structures where the primary concern is the self-weight dead load of the embankment materials themselves. At what thickness does a "pavement" start to look and behave more like an "embankment"? It is dependent on the wheel load of the vehicle.

The vertical stress due to the wheel load can be modeled using the Boussinesq equation [7.18] and compared to the vertical stress imposed by the weight of saturated soil above a point in the subgrade. Where the two curves for the two stress distributions cross, the depth where a pavement becomes more like an embankment can be identified (Figures 7.3 and 7.4). There is a smooth, gradual transition from pavement to embankment, but the depth where the curves cross can serve as an indicator of the change. Note in Figure 7.4 that the depth where designing for pavement behavior is at least as important as designing for embankment behavior is as much as

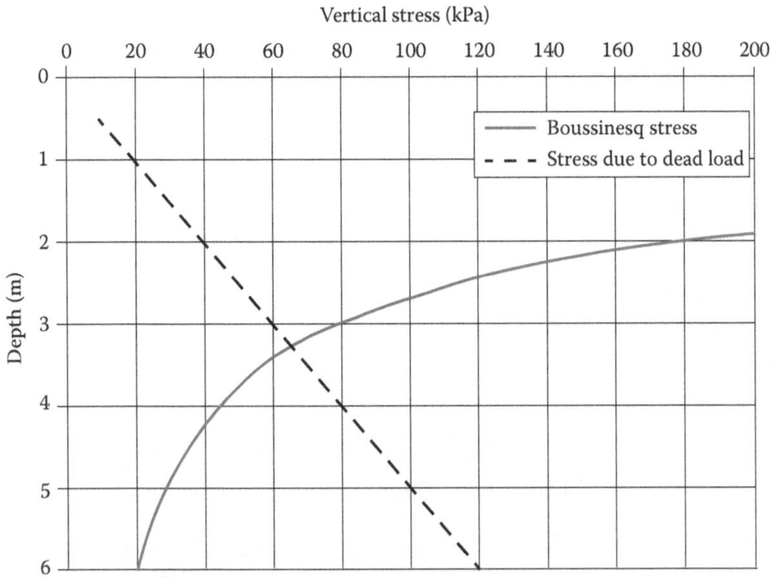

Figure 7.3 Vertical stresses due to live load and dead load, against depth. Live load stresses are for a 300 tonne (660,000 lb) axle with dual tires, dead load stresses are for a soil unit weight of 20 kN/m^3 (130 lb/ft^3).

3.3 m (11 ft) for a 300 tonne (330 ton) axle load. The structure should be designed as a pavement and also checked for adequacy as an embankment (e.g., base bearing capacity, settlement, and slope stability).

7.7.2 Designs from ports and airfields literature

Haul vehicles serving port facilities and large commercial aircraft impose heavy wheel loads on the pavements designed for them, and the pavement design methods for them can be used to design pavements for haul roads, sometimes with some necessary adaptations. Examples of such pavement design methods are presented in Knapton [7.19,7.20], AI [7.21], and Argue [7.22].

7.7.3 CBR chart

Atkinson [7.23] referenced in Tannant and Regensburg [7.24] presents a design chart with base cover thickness plotted against subgrade CBR for a range of *wheel* loads from 12 to 80 tonnes (26,000 to 180,000 lb). The failure criterion is not stated. A procedure is detailed in Tannant and Regensburg [7.24] which facilitates the design of multiple-layer granular pavement designs.

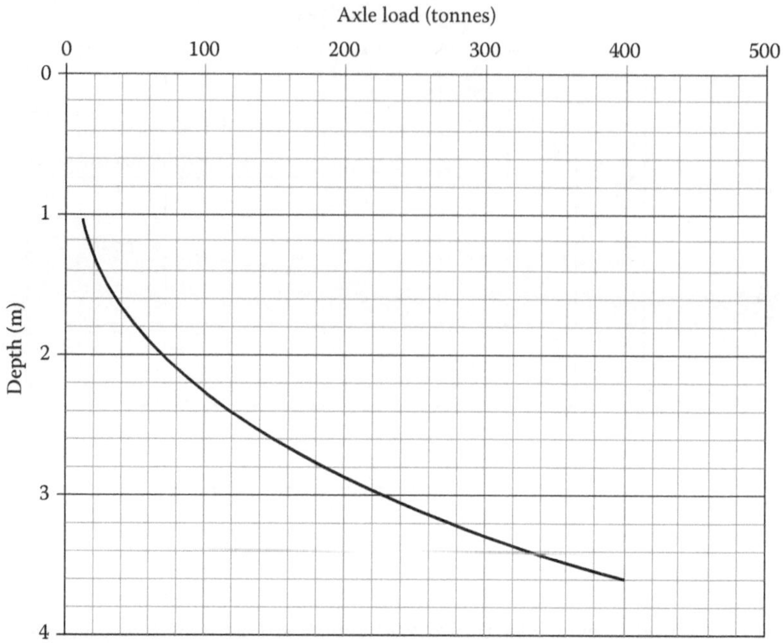

Figure 7.4 Depth where vertical stress due to live load equals vertical stress due to dead load, as a function of axle load.

7.7.4 Design based on "critical strain"

Atkinson's CBR chart [7.23,7.24], although catering to axle loads as great as 160 tonnes (350,000 lb), may still not go far enough for some mining, oil sands, or oil and gas haul road designs. Ultra-heavy dump truck gross vehicle masses as heavy as 600 tonnes (1.3 million lb), with loaded back axles imposing 400 tonnes (880,000 lb) are not uncommon (Figure 7.5). Design based on the *critical strain limit* can be used to design pavements for ultra-heavy haul trucks.

The premise of the critical strain method is that the vertical strain anywhere in the pavement should not exceed a critical limit, which is a function of the number of passes of the design axle [7.24]:

$$\varepsilon_{critical} = \frac{80{,}000}{N^{0.27}} \tag{7.9}$$

where

$\varepsilon_{critical}$ is the critical strain (micro-strain)
N is the number of passes of the design axle load

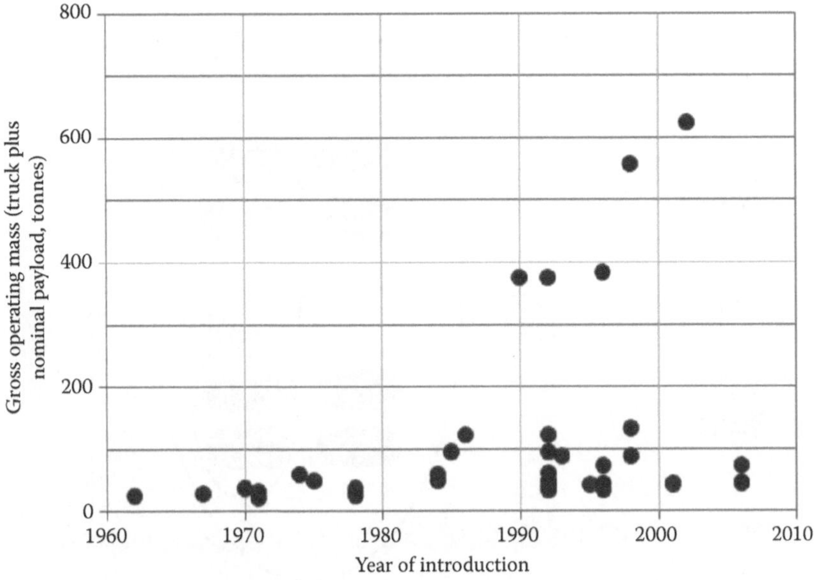

Figure 7.5 Rated operating mass of one manufacturer's off-road mining trucks, by model year of introduction. (From Caterpillar, *Caterpillar Performance Handbook*, 29th edn., Caterpillar Inc., Peoria, IL, pp. 25-70 and 25-71.)

Unfortunately, no design criterion is stated. Morgan et al. [7.26 cited in 7.24] determined that the critical vertical strain at the top of the subgrade should be about 1500 με. Thompson and Visser [7.27, 7.28 cited in 7.24] advocated a maximum strain limit of 2000 με at the top of the road surface. Knapton [7.29 cited in 7.24] developed an equation for heavy loads on docks at container ports; Equation 7.9 is a modified form for roads.

The following steps are carried out in the method:

1. The critical strain limit is calculated (Equation 7.9), usually lying between 1500 and 2000 με.
2. The resilient modulus for each pavement layer is set.
3. Based on experience, initial pavement layer thicknesses are set for a trial pavement design.
4. The vertical strains are calculated at particular points in the pavement cross section.
   a. Finite element analysis or elastic layer analysis is used.
   b. Typical locations where strains are calculated include the mid-depth, top and bottom faces of the pavement layers, and the top of the subgrade, at locations beneath the center of a tire, the center of a dual-tire wheel, the center of an axle, on the centerline of the

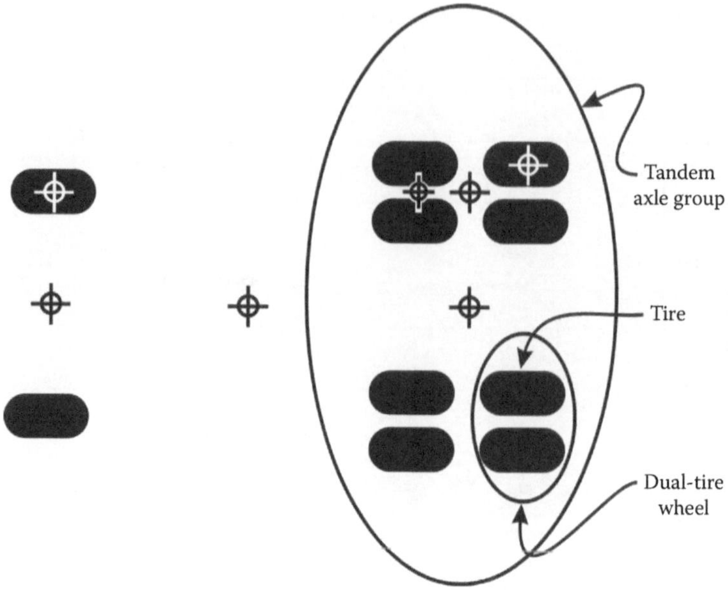

Figure 7.6 Plan view of typical locations where vertical strains are calculated in the critical strain method, for a tandem axle, dual wheel, 3 axle dump truck. Strains are computed at various depths below the locations indicated by symbol. (Not to scale.)

dual-tire wheels between axles in an axle group, at the center of an axle group, and at the center of the vehicle (Figure 7.6).

5. The calculated vertical strains are compared to the critical strain.
   a. If any is greater than the critical strain, the design is unacceptable, and one or more layer thicknesses should be increased, and the process repeated, beginning with Step 3.
   b. If any is much less than the critical strain, the design is acceptable, but inefficient. The layer(s) above the lightly strained layer should be made thinner, and the process repeated beginning with Step 3.
   c. If all calculated strains are equal to or just less than the critical strain, the design is acceptable and efficient.

Care should be taken with the elastic analysis carried out in Step 4 (Douglas et al. [7.30]). Granular materials are incapable of sustaining tensile strains, yet the finite element or layered elastic analysis may erroneously compute significant tensile strains, under-estimating the calculated vertical strains. A "no-tension" finite element or layered elastic analysis should be performed.

REFERENCES

7.1 AASHTO. 1993. *AASHTO Guide or Design of Pavement Structures.* Washington, DC: American Association of State Highway and Transportation Officials (AASHTO).

7.2 Haas, R., Hudson, W.R., and Zaniewski, J. 1994. *Modern Pavement Management.* Malabar, FL: Krieger Publishing Company, 583pp.

7.3 Bowles, J.E. 1992. *Engineering Properties of Soils and the Measurement.* Toronto, Ontario, Canada: McGraw-Hill Inc., 241pp.

7.4 TRL. 1993. A guide to the structural design of bitumen-surfaced roads in tropical and subtropical countries, *Overseas Road Note 31*, 4th edn. Crowthorne, U.K.: Transportation Research Laboratory (TRL), 75pp. http://www.transport-links.org/transport_links/filearea/publications/1_716_Microsoft%20Word%20-%20Overseas%20Road%20Note%2031%20edit1.pdf, visited January 15, 2015.

7.5 Lay, M.G. 2009. *Handbook of Road Technology*, 4th edn. Abingdon, U.K.: Spon Press, 933pp.

7.6 NAASRA. 1979. *Interim Guide to Pavement Thickness Design.* Sydney, New South Wales, Australia: National Association of Australian State Road Authorities (NAASRA).

7.7 Austroads. 2009. *Review of Relationship to Predict Subgrade Modulus from CBR (California Bearing Ratio).* Sydney, New South Wales, Australia: Austroads Inc., 23pp.

7.8 Hajek, J.J., Smith, K.L., Rao, S.P., and Darter, M.I. 2008. Adaptation and verification of AASHTO pavement design guide for Ontario conditions. Toronto, Ontario, Canada: Ministry of Transportation of Ontario, Pavements and Foundation Section. Prepared by ERES Consultants, Applied Research Associates, Inc.

7.9 Smith, K.D., Zimmerman, K.A., and Finn, F.N. 2004. *The AASHO Road Test, Living Legacy for Highway Pavements.* TR News 232: May–June. Washington, DC: US National Academies, Transportation Research Board, pp. 14–24.

7.10 Swift, E. 2011. *The Big Roads: The Untold Story of the Engineers, Visionaries, and Trailblazers Who Created America's Superhighways.* Boston, MA: Houghton Mifflin Harcourt, pp. 198–199.

7.11 MTO. 1990. *Pavement Design and Rehabilitation Manual.* MTO STO-90-01. Toronto, Ontario, Canada: Ministry of Transportation of Ontario (MTO), 277pp.

7.12 MTO. 2004. Material specifications for aggregates—Base, subbase, select subgrade, and backfill material. Ontario Provincial Standard Specification OPSS 1010. Toronto, Ontario, Canada: Ministry of Transportation of Ontario (MTO), 18pp. http://www.raqsb.mto.gov.on.ca/techpubs/ops.nsf/d37f5a16d8174ffa85256d130066857f/702bdff1eaaf4d9e85257b5c006080ac/$FILE/OPSS.MUNI%201010%20Nov%2013.pdf, visited January 13, 2015.

7.13 Austroads. 1992. *Pavement Design, a Guide to the Structural Design of Road Pavements.* Austroads Publication No. AP-17/92. Surrey Hills, New South Wales Australia: Austroads.

7.14 ARRB. 2000. *Unsealed Roads Manual, Guidelines to Good Practice* (Giummarra, G., ed.). Vermont South, Victoria, Australia: ARRB Transport Research Ltd., p. 3.2.

7.15 Hoque, Z. and Jameson, G. 2006. *Pavement Design for Light Traffic: A Supplement to the Austroads Pavement Design Guide*. Sydney, New South Wales, Australia: Austroads Inc., p. 15.

7.16 FHWA. Get to know the new M-E pavement design guide. Focus, July 2006. Washington, DC: US Department of Transportation, Federal Highway Administration (FHWA), 7pp.

7.17 Pierce, L.M. and McGovern, G. 2014. *Implementation of the AASHTO Mechanistic-Empirical Pavement Design Guide and Software*, NCHRP Synthesis 457. Washington, DC: US Transportation Research Board, 71pp.

7.18 Lambe, T.W. and Whitman, R.V. 1969. *Soil Mechanics*, SI Version. Toronto, Ontario, Canada: John Wiley & Sons, Inc., 553pp.

7.19 Knapton. 1996. *The Structural Design of Heavy-Duty Pavements for Ports and Other Industries*, 3rd edn. Leicester, U.K.: Precast Concrete Federation Limited, 45pp.

7.20 Knapton. 2008. *Heavy-Duty Pavements, the Structural Design of Heavy-Duty Pavements for Ports and Other Industries*, 4th edn. Leicester, U.K.: British Precast Concrete Paving and Kerb Association, Precast Concrete Federation Limited, 87pp.

7.21 AI. 1986. *Thickness Design, Asphalt Pavements for Heavy Wheel Loads*, Manual Series 23, 1st edn. College Park, MD: The Asphalt Institute (AI), 69pp.

7.22 Argue, G.H. 2005. *Canadian Airfield Pavement Engineering Reference.* Ottawa, Ontario, Canada: Glen Argue. (May possibly be available from glen.argue@sympatico.ca.)

7.23 Atkinson, T. 1992. Design and layout of haul roads. *Chapter 13.4—SME Mining Engineering Handbook*, 2nd edn. Littleton, CO: Society of Mining, Metallurgy, and Exploration, Inc., pp. 1334–1342.

7.24 Tannant, D.D. and Regensburg, B. 2001. *Guidelines for Mine Haul Road Design*. Edmonton, Canada: School of Mining and Petroleum Engineering, Department of Civil and Environmental Engineering, University of Alberta, 111pp. https://circle.ubc.ca/bitstream/id/90332/Haul_Road_Design_Guidelines.pdf Visited January 17, 2015.

7.25 Caterpillar. undated. *Caterpillar Performance Handbook*, 29th edn. Peoria, IL: Caterpillar Inc., pp. 25-70, 25-71.

7.26 Morgan, J.R., Tucker, J.S., and McInnes, D.B. 1994. A mechanistic design approach for unsealed mine haul roads. *Pavement Design and Performance in Road Construction* 1412:69–81.

7.27 Thompson, R.J. and Visser, A.T. 1997. An introduction to the integrated design of surface mine haul roads. *International Journal of Surface Mining, Reclamation and Environment* 11: 115–120.

7.28 Thompson, R.J. and Visser A.T. 1997. A mechanistic structural design procedure for surface mine haul roads. *International Journal of Surface Mining, Reclamation and Environment* 11: 121–128.

7.29 Knapton, J. 1988. *The Structural Design of Heavy-Duty Pavements for Ports and Other Industries*. London, U.K.: British Ports Federation.

7.30 Douglas, R.A., Laprade, R., and Lawrence, K. 2015. High-strength geotextiles in ultra-heavy-load haul roads. In *Proceedings 11th International Conference on Low-Volume Roads*. Washington, DC: US Transportation Research Board, 13pp.

Chapter 8

Drainage

There are really only three key things to remember about road construction—drainage, drainage, and drainage.

8.1 INTRODUCTION

There are many ways water can flow over, into, and through road structures (Figure 8.1), and the flows can be anything from inconvenient to catastrophic. Water flowing across or along the road surface can be erosive, disrupting the integrity of the surface, and presenting an uncomfortable and potentially dangerous ride for vehicle operators. Water getting into the road structure can weaken the subgrade and pavement layer granular materials sufficiently to cause failure in the form of deep ruts, potholes, and/or punch-outs.

In this chapter, surface drainage is discussed first, then internal drainage. The chapter then winds up with a number of issues related to drainage.

8.2 SURFACE DRAINAGE

If water flows along the road, it usually flows along the wheel path ruts (Figure 8.2). On a sealed pavement (asphalt or surface treated) it will increase the risk of hydroplaning (aquaplaning), where a film of water forms between the road surface and vehicle tires, causing a loss of vehicle control. On a granular pavement (gravel or crushed rock) the water flowing down wheel path ruts may have enough energy to scour the fines out of the surface gravel, unraveling the fabric of the road. The eroded material may eventually find its way to nearby watercourses, damaging aquatic habitat.

To move water off the road quickly, long flat alignments should be avoided. The road surface should be graded with a positive crossfall (camber, crown) sloping from the centerline to the edges of the pavement. A slope of 2%

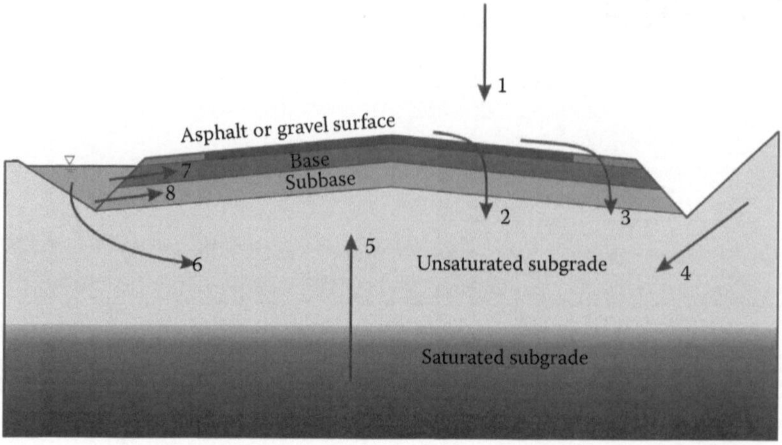

Figure 8.1 The ways water can enter the pavement structure: precipitation can pass through cracks in the surfacing of an asphalt pavement or infiltrate the granular surface layer of a gravel pavement (1); flood waters can pass through cracks in the surfacing of an asphalt pavement or infiltrate the granular surface layer of a gravel pavement (2); water can enter the pavement structure through the granular shoulders (3); water can seep from higher slopes into the pavement structure (4); soil suction can bring water up from the saturated subgrade up into the unsaturated subgrade ("capillary rise") (5); water in inundated drainage ditches can seep into the pavement structure (6), or in the base (7), or subbase (8).

is appropriate for sealed pavements. However, for granular pavements, the crossfall should be 3%–5%, because the rougher surface impedes the movement of water, and because grading operations tend to flatten the crossfall over time.

If you don't look after the water properly, Mother Nature will, and she may do it in a most inconvenient way.

8.2.1 Road cross sections

Roads on flat to rolling ground can have the crossfall established by grading the slope from the centerline to the edges, symmetrically, or if narrow, from one edge of pavement to the other in one plane. When the road is in a sidehill cut, the situation becomes more complicated, with out-sloping and in-sloping cross sections possible, Figure 8.3.

Out-sloping roads encourage surface runoff over the outside edge of pavement at any convenient location, so the runoff is unlikely to be concentrated anywhere. However, there is little control over where the water goes. In-sloping roads are arranged so that the water flows off the road surface to

Figure 8.2 Surface erosion caused by surface water running down the wheel path rather than being directed off the side of the road into a drainage ditch.

a drainage ditch between the road edge and the slope beyond the road edge, and then down along the ditch to a drop structure and a cross-drain culvert which brings it across under the road to an outlet designed with riprap erosion protection. Table 8.1 outlines the differences between out-sloping and in-sloping cross sections.

Whether to use an out-sloping or in-sloping road cross section depends on policy, and on the particular circumstances of the road. However, the decision to build the road one way or the other should be made understanding the advantages and disadvantages of each.

8.2.2 Drainage ditches and cross drains

Drainage ditches are a necessary evil. They control water and carry it away from the road structure. However, they can become choked with vegetation, especially if the adjoining road is travelled by long-haul trucks

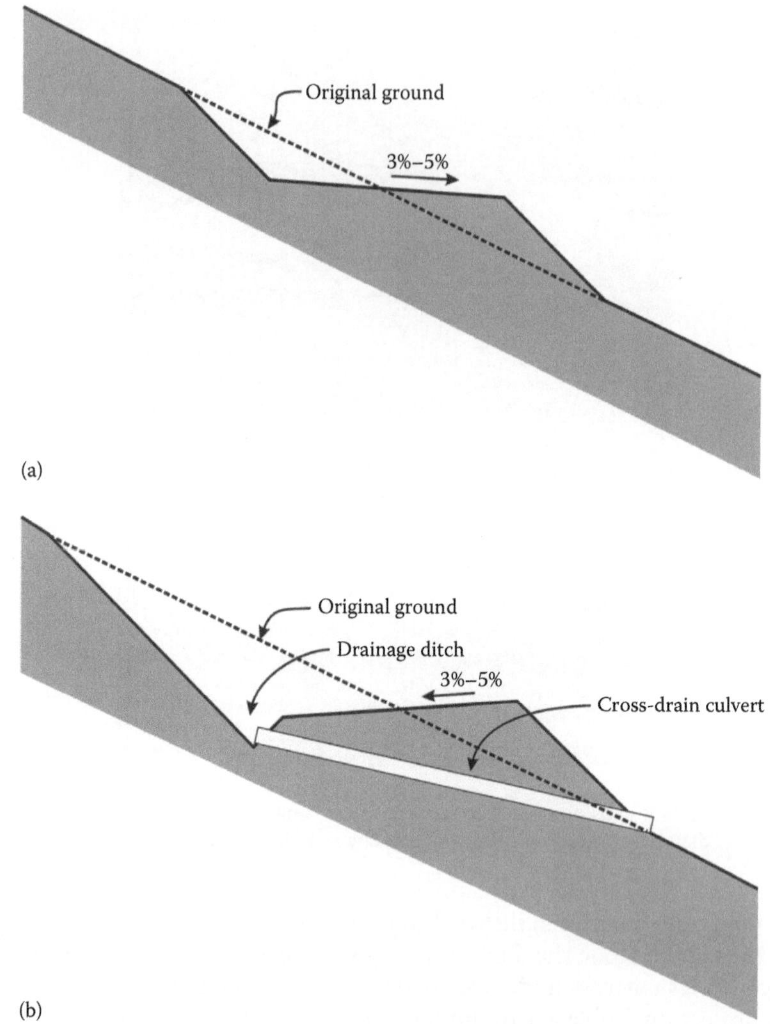

Figure 8.3 Out-sloping (a) and in-sloping (b) cross sections. Note the greater amount of earthworks required to achieve the in-sloped cross section.

that can inadvertently carry and distribute the seeds of virulent invasive species, eventually leading to the necessity to do expensive ditch-clearing work. Not to clear a ditch of vegetation is to run the risk of it overtopping in a storm, or sitting full of water, saturating the pavement structure (Figure 8.1).

Table 8.1 Comparison of out-sloping and in-sloping cross sections

Out-sloping	*In-sloping*
Tends to disperse water	Tends to concentrate water
Water able to leave road surface anywhere along slope	Water must be collected in side ditch
Water is not controlled, may leave road surface in a bad location	Where water leaves the road is controlled
Comparatively small amount of earthwork needed	Comparatively large amount of earthwork needed (drainage ditch, larger cut behind drainage ditch, larger fill below road on downhill side)
No need for cross drains, therefore less expensive	Cross drains needed, adds to expense
No need for cross drains, therefore no maintenance requirement (cleaning the pipes)	Cross drains must be maintained (cleared of debris)
No drainage ditch to erode	Drainage ditch may be eroded, especially on steep slopes (where armoring may be needed), adds to cost
No cross drain outfalls	Cross drains may need erosion protection at outfall
Less safe to drive—road surface tipped toward downhill slope	Safer to drive—road surface tipped toward ditch side

8.2.2.1 Erosion

Ditch bottoms can be damaged by fast flowing water, if the ditch has been dug into erodible material. The erodibility of soils can be determined on the Wischmeier nomograph [8.1], Figure 8.4. The output of the nomograph is an erodibility factor K, ranging from 0 to 1. Highly erodible soils have a K value exceeding 0.6, while relatively nonerodible soils have a value less than about 0.2. Erodibility increases with increasing silt and fine sand content, deceasing organic content, and decreasing permeability.

Another way to check the erodibility of the soil in the ditch bottom is to compare the velocity of the water in the ditch to the maximum tolerable velocity. The velocity of water for uniform flow in an open channel can be estimated using Manning's equation [8.2], Equation 8.1 in metric units, Equation 8.2 in U.S. Customary Units:

$$V = \frac{1}{n}(R^{2/3})(S^{1/2}) \tag{8.1}$$

where

V is the average flow velocity (m/s)
n is a roughness coefficient (unitless)

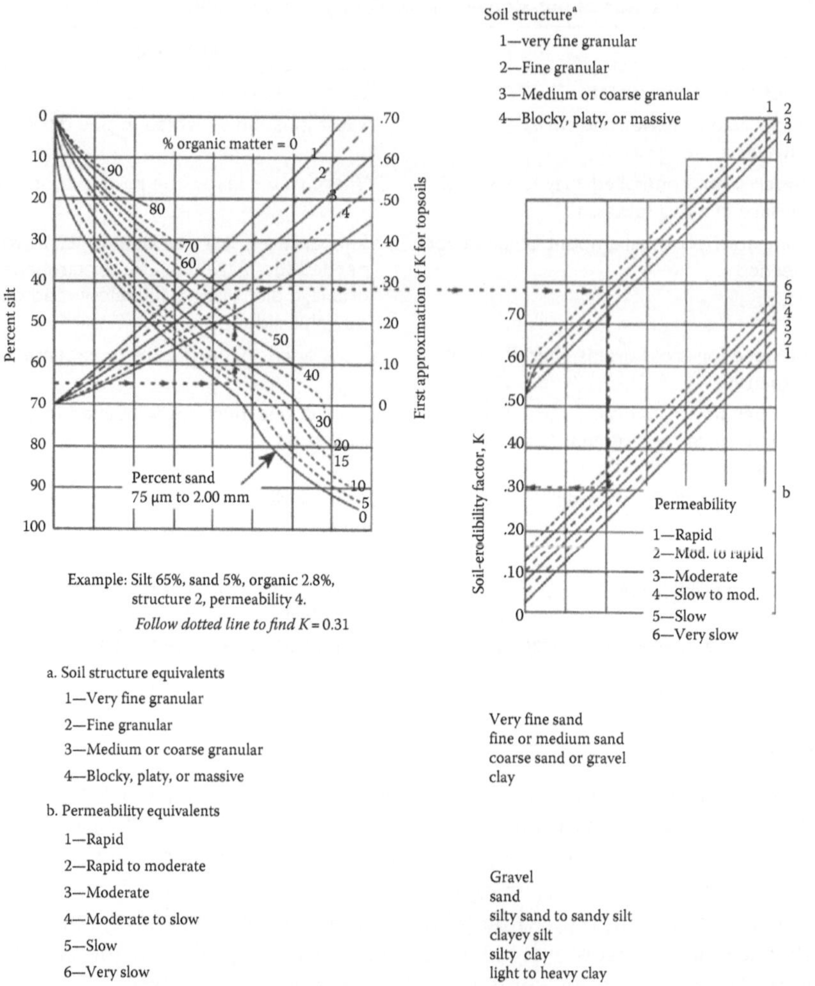

Figure 8.4 The Wischmeier nomograph for determining soil erodibility. (From Wischmeier, W.H. et al., *J. Soil Water Conserv.*, 26(5), 189, 1971 Copyright. Reproduced with permission).

R is the hydraulic radius (m), equal to A/P
A is the cross-sectional area of the flow (m^2)
P is the wetted perimeter (m), the distance along the bottom and sides of the channel in contact with the water
S is the slope of the channel (m/m)

$$V = \frac{1.49}{n}(R^{2/3})(S^{1/2}) \qquad (8.2)$$

where

V is the average flow velocity (ft/s)
n is a roughness coefficient (unitless)
R is the hydraulic radius (ft), equal to A/P
A is the cross-sectional area of the flow (ft^2)
P is the wetted perimeter (ft), the distance along the bottom and sides of the channel in contact with the water
S is the slope of the channel (ft/ft)

The calculated flow can be compared to the maximum value for nonerosive water velocities [8.3], Table 8.2.

Riprap can be used to protect ditch bottoms in extreme cases, when erodible soils or steep ditch slopes cannot be avoided. The required weight of the individual stones is related to the velocity of the water in the ditch [8.4], Figure 8.5. Riprap should be placed on a granular filter (see Section 8.3.8) or suitable geosynthetic to avoid the loss of fines from the soil beneath. Drainage ditches on steep alignments or runs without take-offs into the surrounding ground can be protected with weirs made of hay bales, riprap, or proprietary products.

8.2.2.2 Cross drains

Because the erosive energy the flowing water can have in a ditch is related to the slope of the ditch, steep ditch grades are to be avoided in road

Table 8.2 Maximum values for noneroding water velocities in ditches

	Velocity	
Material	*(m/s)*	*(ft/s)*
Sandy, silty material		
Fine sands and silts	0.40–0.60	1.3–2.0
Clayey material		
Sandy clay	0.50–0.75	1.6–2.5
Clay	0.75–1.00	2.5–3.3
Firm clay	1.00–1.50	3.3–4.9
Gravelly material		
Silty gravel	1.00–1.50	3.3–4.9
Fine gravel	1.50–2.00	4.9–6.5
Coarse gravel	2.00–3.50	6.5–11.5
Other material		
Soft shales	1.50–2.00	4.9–6.5
Rip rap	3.00–4.50	9.8–14.7
Unweathered rock and concrete	4.50–7.50	14.7–24.6

Source: Rodriguez, A.R., del Castillo, H., and Sowers, G.F., *Soil Mechanics in Highway Engineering*, Transtech Publications, Clausthal-Zellerfeld, Federal Republic of Germany, 1988.

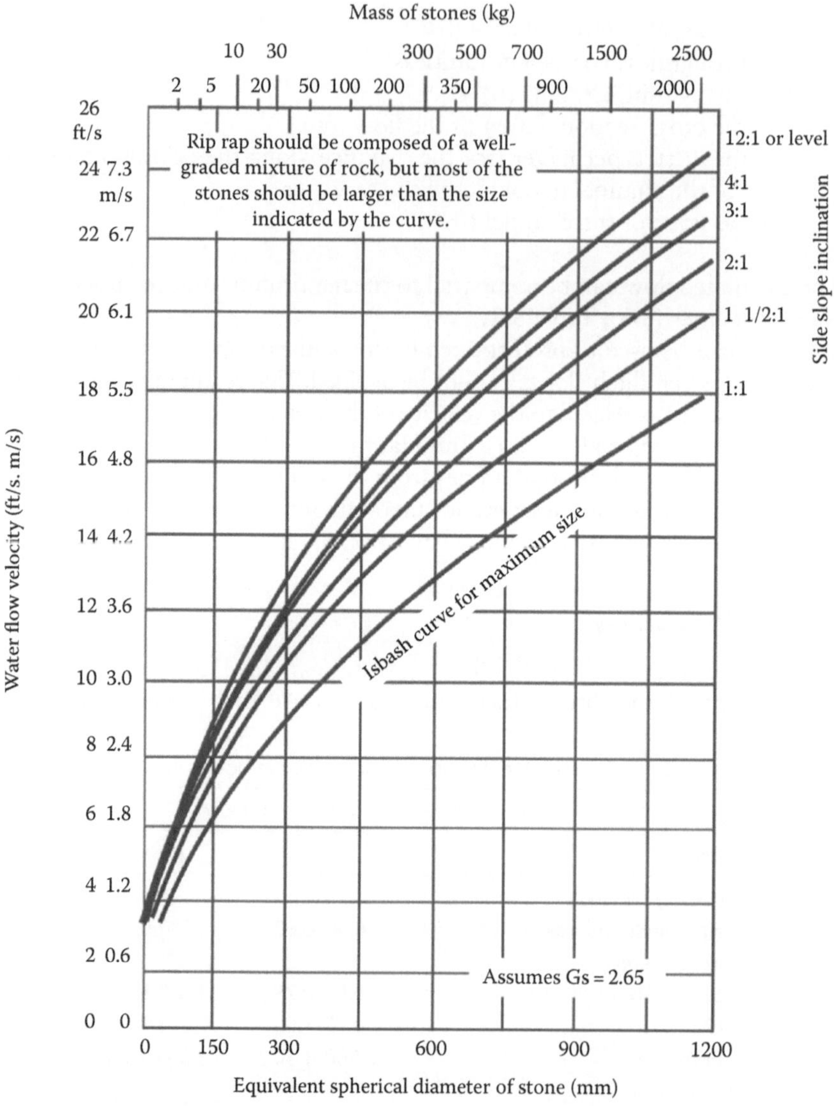

Figure 8.5 Selection of riprap piece size. (FHWA Hydraulic Engineering Circular No. 11, 1978, cited in Keller, G. and Sherar, J., *Low-Volume Roads Engineering, Best Management Practices Field Guide*, U.S. Department of Agriculture Forest Service, Quincy, CA, 2003, 158pp. Used with the permission of Gordon Keller.)

location. Where steep gradients are unavoidable, regularly spaced cross drains should be provided. Cross drains capture the water flowing in the ditch with a drop structure (sump, catch basin, dug-out), and then carry it across in a culvert to the downhill side of the road. They should be spaced

Table 8.3 Recommended spacing for cross drains and rolling dips

	Nonerodible soils		*Erodible soils*	
Road gradient (%)	*(m)*	*(ft)*	*(m)*	*(ft)*
0–3	120	395	75	245
4–6	90	295	50	165
7–9	75	245	40	130
10–12	60	195	35	115
>12	50	165	30	100

Source: Keller, G. and Sherar, J., *Low-Volume Roads Engineering, Best Management Practices Field Guide*, U.S. Department of Agriculture Forest Service, Quincy, CA, 2003, 158pp. Used with the permission of Gordon Keller.

close enough to keep the erosive energy of the water in the drainage ditch to a minimum. Table 8.3 [8.4] shows the recommended spacing for cross drains.

Cross drains should be checked periodically for the accumulation of debris and cleaned out as necessary. In Canada and the northern United States, beavers can do great damage, preferring to plug a culvert to raise the upstream water level enough for them to build their water-bound lodges, rather than going to the much greater effort of building dams. Numerous solutions have been developed to discourage beavers from blocking culverts. Most involve the increase in volume of debris the beaver would need to plug, for instance, a long mesh cage extending outward from the end of the culvert, making plugging the culvert a more onerous task. The ends of cross drains should be clearly staked so that maintenance personnel see them and don't damage the culvert ends, or injure themselves on the culvert ends.

8.2.3 Water bars

A long downhill run of surface water on low-traffic gravel-surfaced road can be disrupted by diverting the water off the road surface at intervals down the hill. Water bars [8.4] can be constructed from materials such as surplus conveyor belting, partially buried in the road structure, with a continuous strip 100 mm tall sticking up, angled at 15° to the cross section, diverting the water off the road. The flexible rubber belting is simply flattened by the tires of vehicles traversing the road, but it springs back up after passage of the vehicle. Providing the road standard is not expected to be very high, the installation of water bars is an appropriate measure for low-speed, gravel-surfaced roads. The recommended spacing of water bars is given in Table 8.4 [8.4].

Table 8.4 Recommended spacing for water bars

	Nonerodible soils		*Erodible soils*	
Road gradient (%)	*(m)*	*(ft)*	*(m)*	*(ft)*
0–5	75	245	40	130
6–10	60	195	30	100
11–15	45	145	20	65
16–20	35	115	15	50
21–30	30	100	12	40
>30	15	50	10	30

Source: Keller, G. and Sherar, J., *Low-Volume Roads Engineering, Best Management Practices Field Guide*, U.S. Department of Agriculture Forest Service, Quincy, CA, 2003, 158pp. Used with the permission of Gordon Keller.

8.2.4 Rolling dips

For low-standard, very low-volume gravel-surfaced roads, rolling dips can be used instead of water bars to interrupt the flow of water down the road, diverting it to the downhill side of the road [8.4] (Figure 8.6). To form a rolling dip, a depression is dug diagonally across the road. Immediately below the depression, the road surface is made to rise slightly. The effect is to stop the water's progress down the road, and divert it to the downhill side of the road. The outlet, of course, should be properly treated with, for example, riprap, to dissipate the concentrated flow coming off the broad-based dip. For rolling dips with high flows, the road surface should also be armored with coarse gravel. The recommended spacing for rolling dips is the same as for cross drains, Table 8.3.

8.2.5 Design

Thus, there is a number of tools to use to deal with the erosive effects of water on steep roads and in the side ditches. On grades up to about 10%, cross drains with culverts or rolling dips are appropriate [8.4]. Between 10% and 15%, closely spaced cross drains with riprap-protected ditches work well. If the gradient exceeds 15%, closely spaced cross drains and riprap-protected ditches are needed, together with armoring of the road surface or a sealed pavement, to protect the surface against erosion. A second look at the road location and alignment would be worthwhile too, to determine if the steep gradient can be avoided.

8.2.6 Beyond the roadway (carriageway)

To some extent, all the measures discussed earlier will change the drainage of the region surrounding the roadway (carriageway). The disruption can

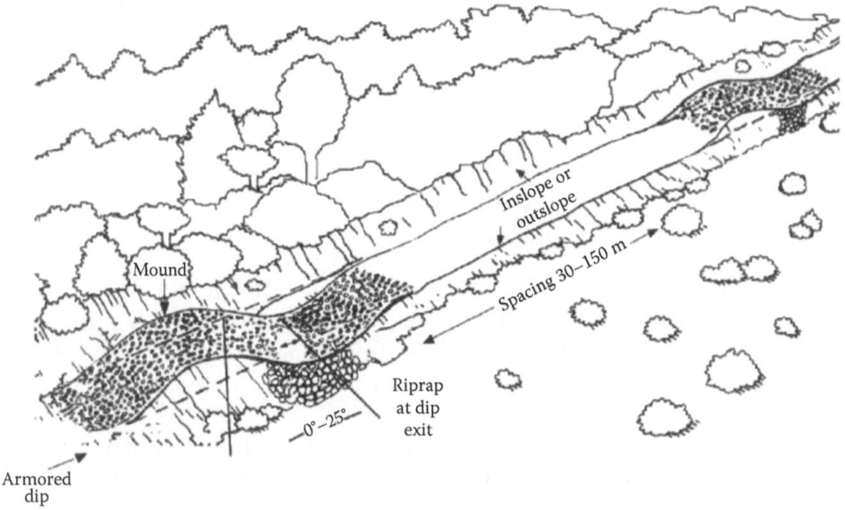

Figure 8.6 Rolling dip. (From Keller, G. and Sherar, J., *Low-Volume Roads Engineering, Best Management Practices Field Guide*, U.S. Department of Agriculture Forest Service, Quincy, CA, 2003, 158pp. Used with the permission of Gordon Keller.)

have far-reaching effects, as in some cases, the road structure can block *regional* flow. Therefore, the impact on the regional surface flows and groundwater flows of the drainage features put in place to improve the drainage of the road must be studied, and responded to.

8.3 INTERNAL DRAINAGE

8.3.1 Importance of internal drainage

As shown in Figure 8.1, there are a number of ways water can enter the road structure (formation) internally. If too much water enters, the subgrade soil may be softened, and/or the pavement layers weakened. If the soils become saturated, there will be a buildup of pore pressure, leading to pumping. The migration and loss of fines in pumping in turn will lead to the formation of potholes. Ultimately, the result will be higher maintenance requirements, a shortened pavement life, or both.

In pavement design (Chapter 7), testing the subgrade strength is recommended. The subgrade strength is a function of the subgrade soil moisture and that can change over time. Permitted enough time, it will come to equilibrium with the boundary conditions, but if those change, the subgrade strength will change, causing a subsequent change in the behavior of the pavement. A common example of this occurs when a granular pavement is sealed after many years. The rate of evaporation decreases significantly, the pavement structure becomes wetter as water coming up from below is no longer evaporated, and the newly sealed road actually becomes softer, with unexpected cracking, rutting, or potholing.

The next sections cover the ways water can move into the pavement structure and solutions to the problems.

8.3.2 Draining water infiltrating through the surface course

Road surface courses are seldom perfectly impervious. Joints and cracks in concrete pavements open up and/or are imperfectly sealed. Thermal cracks form in asphalt pavements. Even if careful attention is paid to specifying a surface layer with fines for granular pavements, water does percolate through.

The base course should be designed to intercept the water coming through the surface course and conduct it to the side ditches or edge drains. The base should be extended under the shoulder paving to "daylight" in the ditch slope, if ditches are chosen as the means of drainage. The ditch invert should be well below the bottom of the base layer so that water doesn't back up into the base when the ditch is flowing full. Otherwise, if edge drains are chosen, the base layer should be in direct contact with the drains.

The following are important points for a base used as drainage blanket, or specific drainage layer placed under the base [8.5]:

- Water should be directed along a path of increasing permeability
- Water should not be trapped in the pavement structure by vertical sags or sudden changes in layer thicknesses
  - This can be a problem where the pavement of one road ties into the pavement of another road at their intersection
- At least 40 mm of base course material should be placed above a drainage course, to protect it from traffic

8.3.3 Draining water rising from the phreatic surface (water table)

The surface tension at the water–air interface in the minute interstices of soils can draw water up from the zone of saturation, into what is called the capillary fringe. The water in a straw rising above the level of the surrounding liquid in a drink illustrates the phenomenon. The suction that results in the soil can be quite large, leading to substantial "capillary rise." The finer the soil is, the greater the rise. Typical values are shown in Table 8.5 [8.6]. Soil suction brings water up from the zone of soil saturation below the phreatic surface, softening the subgrade, and potentially weakening the pavement layers.

Table 8.5 Capillary rise

Soil	Particle size D10 (mm)	Void ratio	Capillary rise (mm)	Capillary rise (in.)
Coarse gravel	0.82	0.27	50	2.1
Sandy gravel	0.2	0.45	280	11
Fine gravel	0.3	0.29	200	7.9
Silty gravel	0.06	0.45	1060	42
Coarse sand	0.11	0.27	820	32
Medium sand	0.02	0.48–0.66	2400	94
Fine sand	0.03	0.36	1660	65
Silt	0.006	0.95–0.93	3590	140

Source: Lambe, T.W. and Whitman, R.V., *Soil Mechanics, SI Version*. 553pp., 1979., Copyright Wiley-VCH Verlag GmbH & Co. KGaA. Reproduced with permission.

One solution to the problem of soil suction bringing water up into the pavement is to install a capillary break. A layer of coarse enough soil, if thick enough, will stop any further capillary rise. The capillary break must be thicker than its own capillary rise.

Another solution is to lower the phreatic surface enough that the capillary fringe above it is well below pavement level. This can be accomplished with the installation of longitudinal edge drains. If deep enough, they will pull the phreatic surface down. A recommendation in the United Kingdom is that the top of the capillary fringe should be at least 1.5 m (5 ft) below subgrade level [8.7]. Edge drains are described in detail in Section 8.3.7.

To ensure the drains are deep enough and designed properly for the conditions, full-scale field trials can be arranged [8.7]. Two trenches can be dug to the planned depth, and a transverse line of boreholes drilled between them. Once the trenches have come to equilibrium with the water at the design level either by draining to a lower level or by pumping, the phreatic surface can be observed in the boreholes. Adjustments in the drainage level are then made on the basis of the depth of the phreatic surface observed in the boreholes.

8.3.4 Draining seepage from hillsides

The cross section of a road in a sidehill cut can intersect a seeping stratum. The hill above the road will deliver water to the road structure through the permeable stratum, softening the subgrade. Solutions use drains to intercept the flow before it gets to the pavement, with drains either above the road or along the pavement edge.

A cut-off drain can be excavated parallel to the centerline of the road, extending to the less permeable stratum below the seeping stratum. It should be sealed at the top to prevent surface water from entering the drain. The surface water might overwhelm the cut-off drain and should be handled with other means. If the seepage layer is thick, it might be more efficient to generally lower the phreatic surface in the vicinity of the road with deep drains.

A common problem is the seepage that accumulates at the transition where a cut road section meets a fill section. A transverse drain can be installed just up-gradient, to cut off the longitudinal seepage.

8.3.5 Drainage at the edge of pavements

As shown by Figure 8.1, water can back up from the side ditch into the subbase or base layers of the pavement, or it may infiltrate the pavement layers from the shoulders. Solutions are to carefully detail the drainage ditch dimensions so that the flood level in the ditch is still below the bottom of the base layer at least, if not the bottom of the subbase layer, or to provide the shoulder with an impermeable surface.

8.3.6 Transfer of vapor

Vapor transfer may bring moisture into the pavement structure. It is usually prevalent in locations where the temperature fluctuates widely during the day [8.7]. Unfortunately, vapor movements are associated with coarse-grained soils, like the sandy gravels and crushed rock used for drainage layers and bases. These are needed for their drainage capacity and cannot be substituted with finer materials that reduce the vapor transfer.

8.3.7 Construction of internal drains

Drains are constructed as trenches filled with free-draining material. The trench is usually lined with a geotextile to filter the fine-grained soil surrounding the trench. Usually a pipe is installed near the bottom of the trench; installing a pipe significantly increases the flow capacity of the drain. The pipes are perforated or slotted, with the perforations or slots facing down to reduce the chance of silt blocking the pipe. In low seepage cases, French drains (drains with no pipes) are sufficient.

The trench is filled with permeable material such as sand, sandy gravel, gravel, or crushed rock. Gap-graded coarse sand, 20 mm crushed rock with 3% or less finer than 75 µm, and open-graded asphalt have been recommended [8.5]. The material must have a permeability >1 mm/s [8.5]. Typical grain size distributions for drain material specified in the United Kingdom [8.7] are given in Table 8.6.

To reduce the amount of fines moving from the soil around the drain into the drain, the trench is lined with geotextile. The geotextile must be chosen to adequately filter the soil, so that there is no loss of mass of the

Table 8.6 Recommended filter materials

BS sieve size (mm)	U.S. sieve size	*Percentage by mass passing a given sieve size* Type A	Type B
63	2 1/2 in.	—	100
37.5	1 1/2 in.	100	85–100
20	3/4 in.	85–100	0–25
10		50–100	0–5
5		35–90	—
1.18	# 16	15–50	—
0.600	# 30	5–35	—
0.150	# 100	0–5	—

Source: O'Flaherty, C.A. (ed.), *Highways, the Location, Design, Construction, and Maintenance of Pavements*, Butterworth-Heinemann, Oxford, U.K., 553pp., 2002.

soil through the movement of fines, leading to subsidence, and no clogging of the free-draining fill in the trench. See Sections 12.2.3 and 12.6.2.8 for a full discussion of geotextile filters.

8.3.8 Granular Filters

Any place water flows through the interface between a fine-grained soil and an appreciably coarser-grained soil, a filter layer is needed. Otherwise, fine particles will be washed from the finer-grained soil, causing a loss of mass and problems like subsidence, and those fine particles will be deposited in the coarser-grained soil, perhaps causing problems like clogging and excess pore pressures. The filter to solve the problems can be a layer of intermediate gradation or a geosynthetic. Geosynthetics are discussed in Chapter 12; soil filters are discussed in the following sections.

In some drainage applications, graded granular filters are appropriate. For example, a drainage blanket in a pavement structure must filter the subbase soil beneath it. If it does not, the fines in the subgrade soil will migrate into the drainage blanket. A material coarse enough to let the water flow through freely, yet fine enough to keep the fine particles of the adjacent soil in place, needs to be specified. The criteria for selecting filter material are related to the grain sizes of the native soil to be filtered, denoted d_i. The grain sizes of the filter are denoted D_j, where i and j are the percentages finer than d and D, respectively. The criteria relate to permeability, infiltration, and segregation [8.5]. If the filter is being used to surround a perforated or slotted pipe in a longitudinal drain, additional requirements to prevent migration of the fines in the filter into the pipe must be met. Table 8.7 provides the criteria.

To filter a fine soil successfully, the grain size distribution curve for the filter soil must fit within the various limits the equations in Table 8.7 set.

Table 8.7 Filter criteria

Aspect		*Criteria*
Permeability		$D_{15}/d_{15} > 5$
Filtration	Unless soil is uniform or well graded	$D_{15}/d_{85} < 5$
		$D_{15}/d_{15} < 20$
		$D_{50}/d_{50} < 25$
	If the soil is uniform (i.e., $d_{60}/d_{10} \leq 1.5$)	$D_{15}/d_{85} < 6$
	If the soil is well graded (i.e., $d_{60}/d_{10} \geq 4$)	$D_{15}/d_{15} < 40$
	If the soil is gap graded	Base filtration criterion on soil finer than the gap
	If the soil has fine layers	Base filtration criterion on the finer soil layers
Segregation during placement		Filter should not be gap graded
		Filter should not contain particles >75 mm
		Filter uniformity coefficient $D_{60}/D_{10} < 20$
Additional criteria for filters placed around perforated or slotted drainage pipes	No more than 5% of the filter material may be finer than 0.075 mm—otherwise fines will migrate into the pipe	
	For perforated pipes	Diameters of holes < D_{85}
	For slotted pipes	Width of slots < 0.83 D_{85}
	For either perforated or slotted pipes	Minimum hole dimension is 0.8 mm

Source: O'Flaherty, C.A. (ed.), *Highways, the Location, Design, Construction, and Maintenance of Pavements*, Butterworth-Heinemann, Oxford, U.K., 553pp., 2002.

Note: *D* refers to filter grain sizes, *d* refers to grain sizes of soil to be filtered, and the subscripts refer to the percentage finer. For example, D_{15} is the filter grain size for which 15% is finer by mass.

For example, suppose a drainage blanket is to be placed under a pavement structure, on a subgrade soil with the grain size distribution curve shown in Figure 8.7. The drainage blanket must filter the subgrade soil. In this case, the permeability, filtration, and segregation criteria of Table 8.7 apply. The following are the inputs:

- $d_{10} = 0.0060$ mm
- $d_{15} = 0.0073$ mm
- $d_{50} = 0.053$ mm
- $d_{60} = 0.08$ mm
- $d_{85} = 0.22$ mm

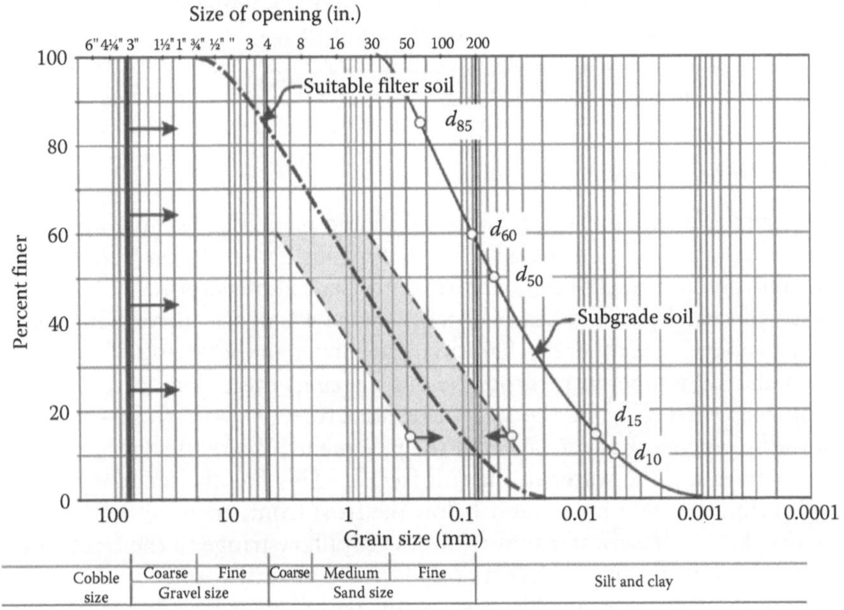

Figure 8.7 Filter design example.

The first step is to determine the permeability criterion: $D_{15} > 5\ d_{15} = 0.037$ mm. Figure 8.7 shows a spot and arrow to indicate this. The next step is to determine the filtration criterion, which will depend on whether the soil is uniform, well graded, gap graded, or none of these. It is well graded, because $d_{60}/d_{10} = 0.08/0.0060 = 13 > 4$. Therefore, the filtration criterion is $D_{15} < 40\ d_{15} = 0.29$ mm. Figure 8.7 shows a spot and an arrow to indicate this. Finally, there are the segregation criteria. The filter should not be gap graded and should have no particles larger than 75 mm. The latter criterion is indicated on Figure 8.7 by a vertical line at a grain size of 75 mm. The last criterion to meet is that the filter should have a uniformity coefficient $D_{60}/D_{10} < 20$. This is indicated in Figure 8.7 by two dashed lines with uniformity coefficient 20 through the spots for the permeability and filtration spots. A material suitable as a filter must have a grain size distribution curve parallel to these lines, run between the permeability and filtration spots, and have all particles smaller than 75 mm. A soil with the dash-dot curve shown in Figure 8.7 would be suitable.

8.4 FROST

There are two aspects to frost action in pavements: the heaving that occurs with frost in the winter, and the subsequent thawing and saturation of pavements in the spring. The heaving and thawing damage pavements and can make driving uncomfortable or even dangerous, however, solutions are available. The following sections will discuss frost heave and frost thaw, and then recommend solutions.

8.4.1 Frost heave

Freezing the water in the pores of a soil should theoretically cause an expansion of about 4%–5% in volume: water expands about 9% on freezing, and the soil's porosity might be about 50%. Fifty percent of 9% is 4.5%. However, pavements can be seen to heave far more than a 4% or 5% increase in volume would cause. Some other mechanism must happen in these cases.

As the air temperature drops below freezing, there is a corresponding drop below freezing in the pavement structure. A *frost front*—the 0°C isotherm—penetrates the layers of the pavement. Where the temperature is below freezing, the water in the soil freezes. This has the tendency to dry and shrink the soil immediately below the frost front, creating suction. The suction forces draw water upward in the capillary fringe to the frost front, where the water delivered freezes on to the ice already formed in the frozen zone of the pavement. Ice "lenses"—pancakes of pure ice—begin to form, continuously thickening as long as they are at the frost front. If the frost front penetrates deeper, the growth of those ice lenses ceases, but others are generated at the frost front lower down. As long as ice lenses grow, the pavement is heaved above.

In the United Kingdom frost is reported to penetrate as much as 0.45 m [8.7]; in Canada, the penetration can be as much as 3 m. Frost heaves of pavement surfaces as much as a foot (0.3 m) or more are not unknown, and heaves inches (centimetres) high are not uncommon. The heaves, and differential heaves, make driving dangerous and can cause severe cracking and distortion of pavement surfaces. The cracking admits more water into the pavement structure and exacerbates the problem.

8.4.1.1 Frost-susceptible soils

The soils susceptible to frost action are those fine enough to generate substantial soil suction forces, yet coarse enough to have a high enough permeability to deliver a large enough quantity of water during the frozen season to build noticeable heaves. Silts and clays with silt seams are susceptible. Sands and gravels do not generate enough suction and so are not susceptible. Clays and silty clays generate high suction forces, but their permeabilities are too low to permit enough water to form ice lenses during the length of winter seen where there are roads.

In the northern hemisphere, this is an important point to note as development pushes northward where freezing temperatures last longer than

Table 8.8 Classification of soils for frost susceptibility

Frost group	*Soil type*	*Percentage finer than 0.020 mm by weight*	*Typical soil types under the unified soil classification system*
F1	Gravelly soils	3 to 10	GW, GP, GW-GM, GP-GM
F2	Gravelly soils	10 to 20	GM, GW-GM, GP-GM
	sands	3 to 15	SW, SP, SM, SW-SM, SP-SM
F3	Gravelly soils	>20	GM, GC
	Sands, except very fine silty sands	>15	SM, SC
	Clays, I_p > 12	—	CL, CH
F4	All silts	—	ML, MH
	Very fine silty sands	>15	SM
	Clays, I_p > 12	—	CL, CL-ML
	Varved clays and other fine-grained, banded sediments	—	CL and ML; CL, ML, and SM; CI, CH, and ML; CL, CH, ML, and SM

Source: Canadian Foundation Engineering Manual/U.S. Army Corps of Engineers. Doré, G. and Zubeck, H.K., *Cold Regions Pavement Engineering*, ASCE Press, New York, 416pp., 2009.

farther south. Where the road structure is frozen longer, finer soils must be considered frost susceptible. Despite their lower permeability, there is more time available for ice lenses to develop within them.

As a guide, nonuniform soils are frost susceptible if more than 3% of the particles are finer than 0.020 mm. Very uniform soils are susceptible if more than 10% of the particles are finer than 0.020 mm [8.7].

In the United Kingdom, it has been noted that cohesive soils are *not* frost susceptible if the plasticity index I_p is greater than 15, or greater than 20 if poorly drained (water level within 0.6 m (2 ft) of pavement). Chalk soils, including crushed chalk, are frost susceptible, and must not be included in road structures subjected to freezing [8.7].

North American practice uses a classification of soils into Frost Groups [8.8]. Soils are assigned to groups F1 to F4 based on their frost susceptibility. Table 8.8 provides the susceptibilities.

8.4.2 Frost thaw

Usually the greater amount of damage takes place during the spring thaw, rather than during the winter heaving. The road structure thaws out from the center top to the bottom and sides. There is a period of time when the frozen lower soil prevents the melt water in the upper soil from escaping, leaving soil saturated or oversaturated and therefore very weak. The surface can break up, the granular layers can be severely weakened, and the pavement deteriorates rapidly. At the least, severe cracking forms; at the worst, the pavement is rendered impassable.

8.4.3 Solutions

Frost action requires all three of the following to be present:

- Freezing temperatures in the pavement
- A source of water
- A frost susceptible soil

Remove one or more of the three, and frost action cannot take place. Therefore, solutions focus on removing one of the requirements.

8.4.3.1 Freezing temperatures in the pavement

Insulation such as rigid foam or brush matting* can be placed in the pavement structure, usually beneath the base or subbase granular layers to insulate the frost-susceptible subgrade soils. With sufficient insulation, the frost front does not penetrate into the susceptible subgrade. Design methods are available to determine the thickness of rigid foam insulation required to protect a frost-susceptible subgrade [8.8].

Another strategy to adopt is to thicken the base and subbase enough that the frost front does not reach down to subgrade level. A thickness equal to or greater than 70% of the expected depth of penetration of the frost front is usually recommended [8.5].

8.4.3.2 Source of water

If the water level in the subgrade is lowered sufficiently, the frost front will not reach the top of the capillary fringe and the frozen zone of the pavement will be deprived of the water needed to form ice lenses. The water level can be lowered by deepening the side ditches or providing subdrains. Clearly some frost-susceptible subgrade soils would require a deepening of the ditches beyond a practical limit.

A second way to deprive the frozen zone of water is to install a capillary break in the susceptible subgrade soil. Insertion of a layer of non-frost-susceptible soil such as clean gravel will interrupt the capillary rise. The capillary break has to be located below the expected maximum depth of penetration of the frost front. The method has the advantage of using

* In forest road construction, tree limbs and tops are placed on the subgrade to a loose depth of a metre (3 ft) or more, tramped down with a bulldozer or excavator, then topped with the granular pavement layers. The construction is called brush matting.

the same subgrade soil. It has the disadvantage of requiring a substantial amount of earthwork.

8.4.3.3 Frost-susceptible soil

A simple expedient is to remove the frost-susceptible subgrade soil and replace it with soil that is not susceptible. Usually replacement down to a depth of about 70% of the expected maximum frost penetration is sufficient [8.5]. While it is simple and effective, the method has the disadvantage of requiring the excavation and wastage of a substantial amount of subgrade soil.

8.5 PERMAFROST

As northern hemisphere development pushes northward, more and more roads are encountering areas of permafrost and discontinuous permafrost. The subject is outside the scope of this book; however, attention is drawn to an excellent reference, *Cold Regions Pavement Engineering* [8.9].

8.6 PUMPING AND POTHOLES

Lay [8.5] explains the phenomenon of pumping as the movement of water, which is incompressible, through cracks in the road surface when the instantaneous pressure of a passing wheel is applied. The water carries the fines of the pavement materials with it, causing a loss of mass in those layers, the formation of voids, and the eventual local collapse of the pavement surface, seen as potholes.

Potholes can also be formed by the upward pressure of water in a saturated pavement either side of the contact between tire and road surface, as the wheel load passes by [8.5]. The pressure exerted by the tire pressurizes the pore water in the saturated pavement, and it finds relief by cracking the surface course of the pavement. Eventually potholes form.

Granular surfaced pavements can have regions of segregation, resulting in pockets of coarse aggregate lacking fines. The pockets fill with water, which is pressurized and pumped with the passage of wheels. The clean coarse aggregate is easily moved, leaving behind a pothole.

The common factor in these pothole-producing mechanisms is saturated pavement materials. If properly drained, there is less risk of potholes forming.

REFERENCES

8.1 Wischmeier, W.H., Johnson, C.B., and Cross, B.V. 1971. A soil erodibility nomograph for farmland and construction sites. *Journal of Soil and Water Conservation* 26(5):189–193.

8.2 Chow, V.T. 1959. *Open-Channel Hydraulics*. Toronto, Ontario, Canada: McGraw-Hill Book Company, 680pp.

8.3 Rodriguez, A.R., del Castillo, H., and Sowers, G.F. 1988. *Soil Mechanics in Highway Engineering*. Clausthal-Zellerfeld, Federal Republic of Germany: Transtech Publications, 843pp.

8.4 Keller, G. and Sherar, J. 2003. *Low-Volume Roads Engineering, Best Management Practices Field Guide*. Quincy, CA: U.S. Department of Agriculture Forest Service, 158pp.

8.5 Lay, M.G. 2009. *Handbook of Road Technology*, 4th edn. Abingdon, U.K.: Spon Press, 933pp.

8.6 Lambe, T.W. and Whitman, R.V. 1979. *Soil Mechanics, SI Version*. New York: John Wiley & Sons, 553pp.

8.7 O'Flaherty, C.A. (ed.). 2002. *Highways, the Location, Design, Construction, and Maintenance of Pavements*. Oxford, U.K.: Butterworth-Heinemann, 553pp.

8.8 CGS. 2007. *Canadian Foundation Engineering Manual*. Richmond, British Columbia, Canada: Canadian Geotechnical Society (CGS), 488pp.

8.9 Doré, G. and Zubeck, H.K. 2009. *Cold Regions Pavement Engineering*. New York: ASCE Press, 416pp.

Chapter 9

Soil erosion and sedimentation control

9.1 SOIL EROSION PROCESS

It is estimated that hundreds of billions of ton(ne)s of topsoil are lost due to erosion annually [9.1] and that the loss of soil outstrips its regeneration by 10–40 times [9.2]. It is partly responsible for the degradation of aquatic habitat and the deterioration of drinking water supplies. A significant portion of the erosion is caused by poor low-volume road design and construction. However, it is a problem that can be mitigated with good design and construction.

The soil erosion process begins during a storm with the bombardment of the soil surface by each rain drop during a storm. A stunning photograph [9.3] shows the moment of impact of a drop of rain on the surface of the soil. In the photo, soil particles can be seen to be detached by the splashing of the rain drop, starting them moving downslope with the water's encouragement. If left unchecked, rain drop erosion can coalesce into sheet erosion, then rill erosion, gully erosion, and channel erosion (Figure 9.1).

Erosion is dependent on rainfall characteristics, including rainfall intensity, duration, and droplet size. Thinking of the soil being bombarded by rain drops, and the energy of that bombardment, it is clear that more intense storms (mm/h or in./h), longer lasting rainfall events, and/or larger droplets will input more energy, resulting in greater amounts of soil moved.

Erosion is also dependent on soil characteristics, including soil texture, organic content, structure, and permeability. The particles of fine-grained soils such as silts and silty fine sands are easily detached; coarser-grained soils require more energy than the rain drop impact can supply, while finer-grained soils such as silty clays and clays have enough cohesion to resist the impact of the rain drops, keeping the particles in place. A high enough organic content will have the same effect, "gluing" the soil particles in place. Soils with cracks and fissures, or layers of coarser material, will permit the rain water to infiltrate rather than run off, reducing the erosion. The same is true of soils with high permeability—the water will infiltrate rather than run off, thus reducing the potential for erosion.

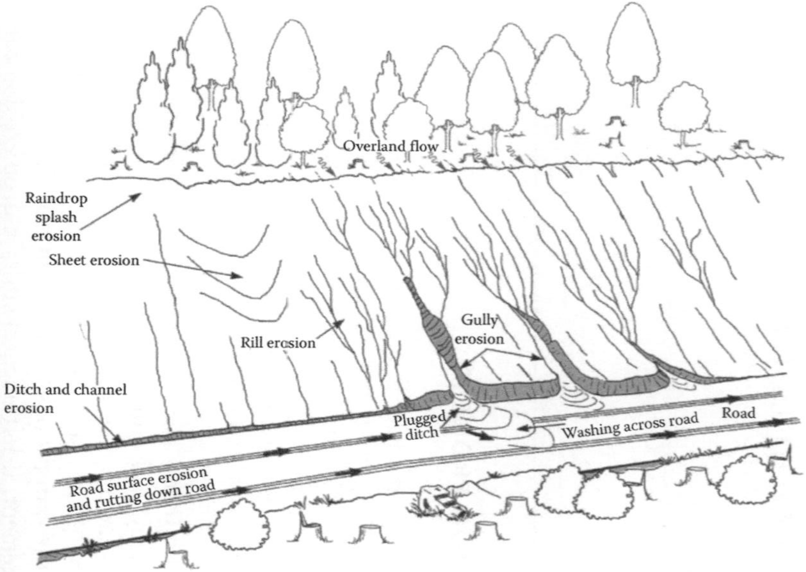

Figure 9.1 Types of erosion. (Used with permission of Gordon Keller.)

The erosivity of soils is discussed further in Chapter 8. In the context of ditch erosion, equations and a nomograph predicting erosivity are presented there.

9.2 WHICH IS PREFERABLE, EROSION CONTROL OR SEDIMENTATION CONTROL?

Where should the effort be concentrated?

Erosion control is the more desirable, because with it, the problem can be caught at the source. Sedimentation control, while also desirable, actually chases the problem after it has occurred. Damage is done along the path of erosion, and the sedimentation can never quite be completely eliminated.

In the real world, however, both are needed.

9.3 EROSION CONTROL

Strategies to adopt to control erosion include the following:

- Shielding the soil surface from the impact of rain drops
- Flattening slopes to reduce the energy available in the flowing water to initiate sheet, rill, gully, and channel erosion
- Slowing the flow of water, thus reducing its erosive energy, by installing traps and/or weirs in the drainage path
- Keeping the sediment on site

To shield the soil surface, construction should be organized so that only the most immediately necessary area of the project is left stripped bare of vegetation. Stripping in stages is preferable to stripping the whole site, because the length of time bare earth is exposed to rain storms is reduced. Once construction is complete, finished slopes and bare earth should be revegetated and/or mulched with straw, hay, or hydro-seeding, a mixture of fiber mulch, seed, fertilizer, and water-sprayed or water-jetted onto the exposed soil surface.

Slopes should be made as flat as possible. Reducing or eliminating high cuts and fills will reduce the energy of surface water flowing over the slope, reducing the prospect of erosion.

Sediment can be retained on site by traps and/or weirs installed in drainage courses. Ditches can be constructed with traps dug at intervals. Where it encounters a trap or weir, the velocity of the water decreases, and, lacking the energy to keep the sediment in suspension, allows it to drop out.

Denuding the soil of its protective cover of vegetation is unavoidable in road construction work, so another strategy is to trap the sediment on site by using *silt fences*. These are temporary fences installed across the path the sediment-laden surface water will take (Figure 9.2). The silt fence filters

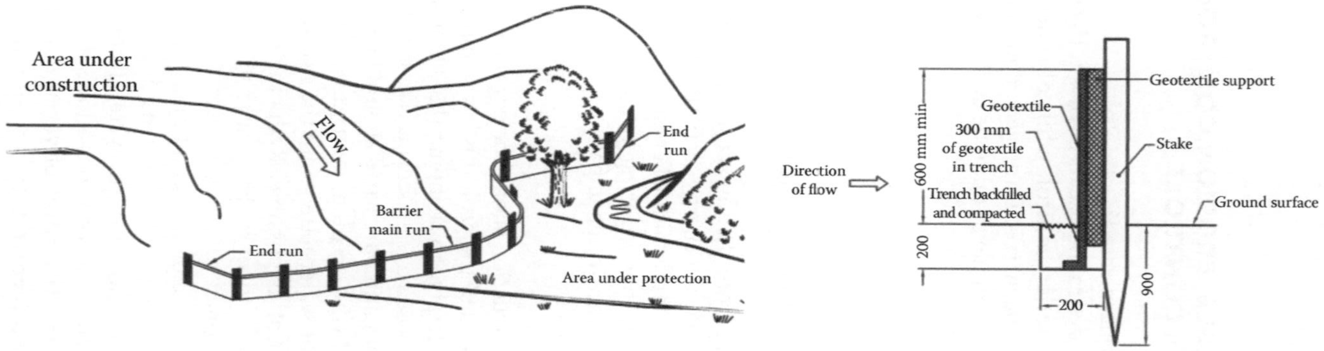

Figure 9.2 Silt fence details. (OPSD, Ontario provincial standards for roads and public works: 219.130 Heavy duty silt fence barrier. Ontario Provincial Government, Ontario Provincial Standard Drawings (OPSD). http://www.raqsb.mto.gov.on.ca/techpubs/ops.nsf/20746bdcd064df1f85256d130066857e/c3874c7b393cb629852572c600647bba?OpenDocument, November 24, visited 2013, 2006. © Queens's Printer for Ontario, 2015. Reproduced with permission.)

the soil particles out, but permits the water to continue to flow by. The fences are constructed from purpose-made geotextile fabrics, wired onto steel or wooden stakes.

Numerous geotextiles are available for silt fencing. The three important criteria to consider when specifying silt fence geotextiles are as follows:

- Durability
- The opening size relative to the size of the soil particles to be filtered
- The flow capacity through the geotextile

Reference can be made to the American standard in ASTM D6461-99(2007)e2 [9.4].

9.4 SEDIMENTATION CONTROL

The driver for sedimentation control is the desire to keep soil particles out of watercourses, to avoid the degradation of aquatic habitat and drinking water sources. If flowing, sediment-laden water can be slowed down, the sediment it carries will settle out and not be carried further down-gradient or downstream. This can be accomplished by flattening slopes.

Sedimentation control measures pertaining to road surfaces are discussed in Chapter 8. That chapter also discusses the armoring of ditches to provide erosion and sediment control.

To provide sedimentation control by slowing water flowing in ditches, weirs made with hay bales, rip rap piles, or proprietary products (Figure 9.3) can be used.

Sediment in roadside ditches is prone to flowing into watercourses at watercourse crossings. Crossings should be designed with ditch "takeoffs," which divert the ditches into the land surrounding the road. The water in the ditch is dispersed with its sediment onto the land, rather than being conveyed to the watercourse. In addition, "no grub zones" should be instituted either side of the watercourse at watercourse crossings. Within a zone of at least 30 m (100 ft) either side of a watercourse at a crossing, no grubbing is permitted for the construction of the road. Low vegetation and roots are left intact after any trees and shrubs within the footprint of the road are cleared, providing a zone where sediment detached by the construction of the road is trapped before reaching the watercourse.

9.5 FOLLOW-UP

As an important part of a road construction project, the surrounding watercourses should be examined to determine their capacity to accommodate

Figure 9.3 Plastic weirs placed in ditch with steep grade, to reduce water velocity to settle out sediment.

increased flows caused directly or indirectly by the road construction project. The new construction will alter and concentrate flows.

Erosion and sediment control facilities must be maintained. They do deteriorate over time and can be damaged by construction activities or vandals, but they are important to the care of the surface water conditions on neighboring properties. The facilities should be inspected regularly, particularly at the peak time of rainfall events, or during the monsoon season, or when there is seasonal snowmelt upstream. Poorly maintained facilities can actually augment the amount of erosion and sedimentation, by releasing sediment stored in the sediment traps and weirs to the watercourse. The traps and weirs should be inspected, together with the silt fences, and accumulations of sediment. Torn or dislodged fencing should be reinstated or patched, and reanchored below ground surface.

REFERENCES

9.1 Muir, P.S. 2014. Course notes, Oregon State University: BI301 Human impacts on ecosystems. Erosion from inappropriate agricultural practices on crop lands. http://people.oregonstate.edu/~muirp/erosion.htm, visited December 1, 2013.

9.2 Lang, S.S. 2006. 'Slow insidious' soil erosion threatens human health and welfare as wells as the environment, Cornell study asserts. Cornell Chronicle, March 20, 2006. Ithaca, NY: Cornell University. http://www.news.cornell.edu/stories/2006/03/slow-insidious-soil-erosion-threatens-human-health-and-welfare, visited December 1, 2013.

9.3 Photograph of moment of impact of rain drop on soil surface: Swac.umn.edu, https://www.google.ca/search?hl=en&site=imghp&tbm=isch&source=hp&biw=1152&bih=633&q=rain+drop+erosion&oq=rain+drop+erosion&gs_l=img.3..0i10i24.3643.7448.0.8349.17.17.0.0.0.0.101.1465.16j1.17.0....0...1ac.1.32.img..0.17.1464.xEHi509OrbA, visited November 24, 2013.

9.4 ASTM. 2007. ASTM D6461–9(2007)E2: Standard specification for silt fence materials. American Society for Testing and Materials (ASTM). http://webstore.ansi.org/FindStandards.aspx?SearchString=D6461&SearchOption=0&PageNum=0& SearchTermsArray=null%7cD6461%7cnul, visited November 24, 2013.

9.5 OPSD. 2006. Ontario provincial standards for roads and public works: 219.130 Heavy duty silt fence barrier. Ontario Provincial Government, Ontario Provincial Standard Drawings (OPSD). http://www.raqsb.mto.gov.on.ca/techpubs/ops.nsf/2 0746bdcd064df1f85256d130066857e/c3874c7b393cb629852572c600647bba?OpenDocument, visited November 24, 2013.

Chapter 10

Watercourse crossings

10.1 INTRODUCTION

Watercourse crossings such as fords, culverts, and bridges are a significant expense, and can be a source of problems, on low-volume networks. Good design can alleviate many of the problems. A watercourse crossing location should be well chosen, an appropriate type crossing selected, and the opening sized correctly. The infrastructure should be installed properly and then well maintained. This chapter covers all of these aspects.

10.2 HYDROLOGY

Watercourse crossing structures need to be properly sized to carry the peak flow rate of the design storm. The rate at which the water is delivered to the structure is determined by the hydrology of the situation; the equations of hydrology convert the characteristics of a precipitation event and the terrain on which the precipitation falls into the peak flow rate at the location of the watercourse crossing structure.

Keller and Sherar [10.1] list a number of equations and approaches to use to estimate the flow, depending on the size of the watershed (Table 10.1). For small watersheds (<120 ha, <300 acres), the Rational Formula is commonly used, Equation 10.1 [10.1] or 10.2:

Metric units:

$$Q = \frac{CiA}{362} \tag{10.1}$$

where

Q is the peak flow rate (m^3/s)
C is a runoff coefficient, reflecting watershed characteristics (unitless)

Table 10.1 Design analysis methods for various watershed sizes

Size	Area (ha)	Area (acres)	Analysis method
Small	to 120	to 300	Rational method, Talbot method, local experience
Medium	to 4000	to 10,000	Regression analysis, high water marks and Manning's equation, local experience
Large	over 4000	over 10,000	Gauging data, high water marks, statistical or regression methods

Source: Keller, G. and Sherar, J., *Low-Volume Roads Engineering, Best Management Practices Field Guide*, U.S. Department of Agriculture Forest Service, Plumas National Forest, CA, 158pp., 2003. Used with the permission of Gordon Keller.

i is the average rainfall intensity for the specified return period storm (mm/h)
A is the area of the watershed feeding the culvert (ha)

U.S. customary units:

$$Q = CiA \tag{10.2}$$

where
Q is the peak flow rate (ft^3/s)
C is a runoff coefficient, reflecting watershed characteristics (unitless)
i is the average rainfall intensity for the specified return period storm (in./h)
A is the area of the watershed feeding the culvert (acres)

C reflects the characteristics of the watershed, ranging from wooded areas to paved surfaces. Table 10.2 [10.1] provides guidance on the selection of C. The larger C is, the greater the peak flow that will have to be passed by the watercourse crossing structure. Scanning Table 10.2, it can be seen that the peak flow is greater the more impervious the terrain is. The peak flow is greater if the precipitation falls on a surface that allows little infiltration and speeds the water downhill to the structure.

10.3 FORDS AND LOW-WATER CROSSINGS

A ford is a dip in the road which goes through a stream rather than passes over it. The road surface is submerged at least some of the time, but most of the time the water depth is shallow enough that the passage of vehicles is not impeded [10.2]. The maximum water depths for cars and trucks are about 0.25 and 0.45 m (10 and 18 in.), respectively, provided there are no cobbles or boulders in the flow.

Table 10.2 Values of *C* for rational method

Land use		*C*
Agriculture	Bare soil	0.20–0.60
	Cultivated fields (sandy)	0.20–0.40
	Cultivated fields (clayey)	0.30–0.50
Grass	Turf, meadow	0.10–0.40
	Steep grassed areas	0.50–0.70
Woodland/forest	Woodland areas with level ground	0.05–0.25
	Forested areas with steep slopes	0.15–0.40
	Bare areas, steep and rocky	0.50–0.90
Roads	Asphalt pavement	0.80–0.90
	Concrete or cobblestone pavement	0.60–0.85
	Gravel surface	0.40–0.80
	Native soil surface	0.30–0.80
Urban areas	Residential, flat	0.40–0.55
	Residential, moderately steep	0.50–0.65
	Commercial or downtown	0.70–0.95

Source: Keller, G. and Sherar, J., *Low-Volume Roads Engineering, Best Management Practices Field Guide*, U.S. Department of Agriculture Forest Service, Plumas National Forest, CA, 158pp., 2003. Used with the permission of Gordon Keller.

10.3.1 Where fords can be used

A ford can be an appropriate choice for a watercourse crossing where there is usually no flow, or a low flow of water, and a low volume of traffic to be accommodated. The traffic has to be able to tolerate occasional periods of suspension of service, when the water level is too high, and or there is a substantial amount of debris passing the ford. Wide flat valleys are the best settings for fords.

A ford can be an acceptable alternative in the following circumstances [10.3]:

- Where a high level approach with embankments would not be environmentally acceptable
- Where a small bridge with embankments would constrict the flow
- In tropical areas where the watercourse usually carries little or no flow but has occasional periods of high flow of short duration
- In cold weather areas where the watercourse usually carries little or no flow, but carries a very high flow during the spring runoff season
- In mountainous terrain where there are unstable channels over alluvial fans—fords represent a low-cost solution to these risky watercourse crossings

10.3.2 Advantages and disadvantages

Fords have numerous advantages compared to bridges and culverts [10.1]:

- Fords are not susceptible to plugging with debris or being blocked by beavers
- Fords are usually less expensive to build and maintain than bridges or culverts, requiring less fill than culverts
- *Vented* fords (Figure 10.1) which have a low running surface which can be overtopped, built on one or more culverts that carry low flows, can be better than fords, as the vehicles are kept out of the water. If overtopped, traffic is temporarily suspended
- Vented fords are good preservers of aquatic habitat—vehicles never drive on the watercourse bottom
- Fords can be designed as broad-crested weirs, thus not being sensitive to large variations in flow—a small increase in water depth over the length of the ford results in a large increase in flow

Fords have a few disadvantages. They may not be environmentally acceptable, if driving on the watercourse bottom will damage aquatic habitat. Fords may inhibit fish passage. Vented fords may be prone to clogging with debris brought down by the watercourse in severe storms. Where traffic volumes or stream velocity is high, fords can be dangerous.

Figure 10.1 Vented fords promote fish passage and permit traffic to stay out of the water. However, they can be prone to clogging with debris or ice if small diameter culverts are used, or if the size and amount of debris or ice is large. (Used with permission of Gordon Keller.)

10.3.3 Fords: Good practice

It is good practice to amour the driving surface with rock which will not be carried away by the water flow. The coarse rock of the ford can be chinked (blinded) with finer gravel to create a smooth running surface. Fords can be improved structures with rock-filled gabion baskets or concrete.

The driving surface of vented fords, which will be overtopped some of the time, should be constructed from a concrete slab or rock amour which extends a safe distance above the design peak water level.

The entire ford or vented ford structure should be protected against scour, with appropriate wing walls, slabs, gabions, and/or bank protection. The downstream edge of the ford is particularly vulnerable to scour and should be constructed with a scour-resistant apron and energy-dissipating features (concrete blocks, boulders, gabions).

Fords should be located where stream banks are low and the channel well defined. They should be founded on scour-resistant strata (bedrock, coarse gravel), below the anticipated maximum depth of scour, protected with rip rap or gabion baskets. Table 8.2 gives estimates of the maximum noneroding water velocities for various materials.

Robust markers should be placed to indicate the edges of the ford to drivers. A depth gauge should be installed, together with a notice of the highest water level for safe passage. Signs warning of the presence of the ford and the possibility that it may be out of service and the road blocked should be placed well up the road on both ends of the ford. If the watercourse carries cobbles and boulders or debris during periods of high flow, warnings should be posted.

10.4 CULVERTS

Besides serving as cross drains (Chapter 8), culverts are used to carry watercourses under LVR. Culverts come in a wide variety of materials, constructions, and shapes [10.4]:

- Materials—galvanized steel, concrete, plastic
- Constructions, galvanized steel—single-piece pipe, bolted multiplate
- Shapes, galvanized steel—round, vertical ellipse, pipe arch, horizontal ellipse, arch (half-circle, reentrant, low-rise), underpass (standard, pear-shaped, horseshoe-shaped)
- Shapes, concrete—round, elliptical, arch, box (open- and closed bottom)

There has been a trend toward the specification of open-bottomed arch culverts. These more closely model the bottom of the natural channel and are thus friendlier to aquatic organisms. Larger sized culverts are also being specified, to reduce peak flow velocities and to reduce the risk of washout.

In the following sections, emphasis is placed on cylindrical corrugated steel pipe culverts, but the principles generally apply to corrugated culverts of other shapes and bottom conditions, as well as culverts fabricated in plastic and concrete.

10.4.1 Culvert sizing

Culverts are sized to be capable of passing the peak design flow, Q m^3/s (ft^3/s). The design is said to be balanced if the culvert can pass the flow resulting from a storm with a 10-year return period with no head above the culvert obvert (top of pipe), and a 100-year return period storm with an acceptable freeboard on the crest of the embankment (Figure 10.2) [10.4]. A 10-year return period storm has a statistical chance of occurring every 10 years. Some agencies design for single return period storms, for example, 25 or 50 years [10.1].

The flow Q calculated in either Equation 10.1 or 10.2 must pass through the culvert safely. Depending on circumstances, the flow through the culvert will be controlled by the inlet conditions or the outlet conditions. If under *inlet* control, the shape of the inlet (e.g., projecting from the fill, mitred to conform to the fill slope, with or without headwalls, equipped with a beveled ring, etc.), the headwater depth, and the cross-sectional area will control the inlet flow capacity of the culvert. If under *outlet* control, the capacity of the culvert will be controlled by the height of the tail water in the outlet channel and the slope, roughness, and length of the culvert [10.4]. Outlet control, where the tail water is high above the culvert obvert, is to be avoided, because in this circumstance, the bedding and backfill become saturated and weakened.

Derived from Bernoulli's Equation, nomographs exist for various shapes of culverts; Figure 10.3 shows a nomograph for cylindrical corrugated steel culverts.

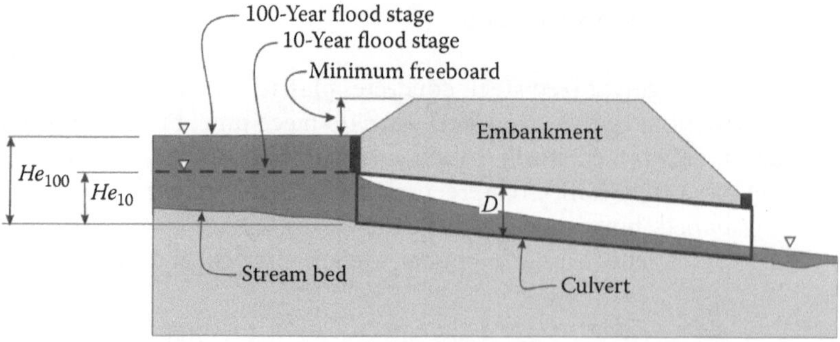

Figure 10.2 Cross section of road and culvert showing conditions for balanced culvert design. (Not to scale.)

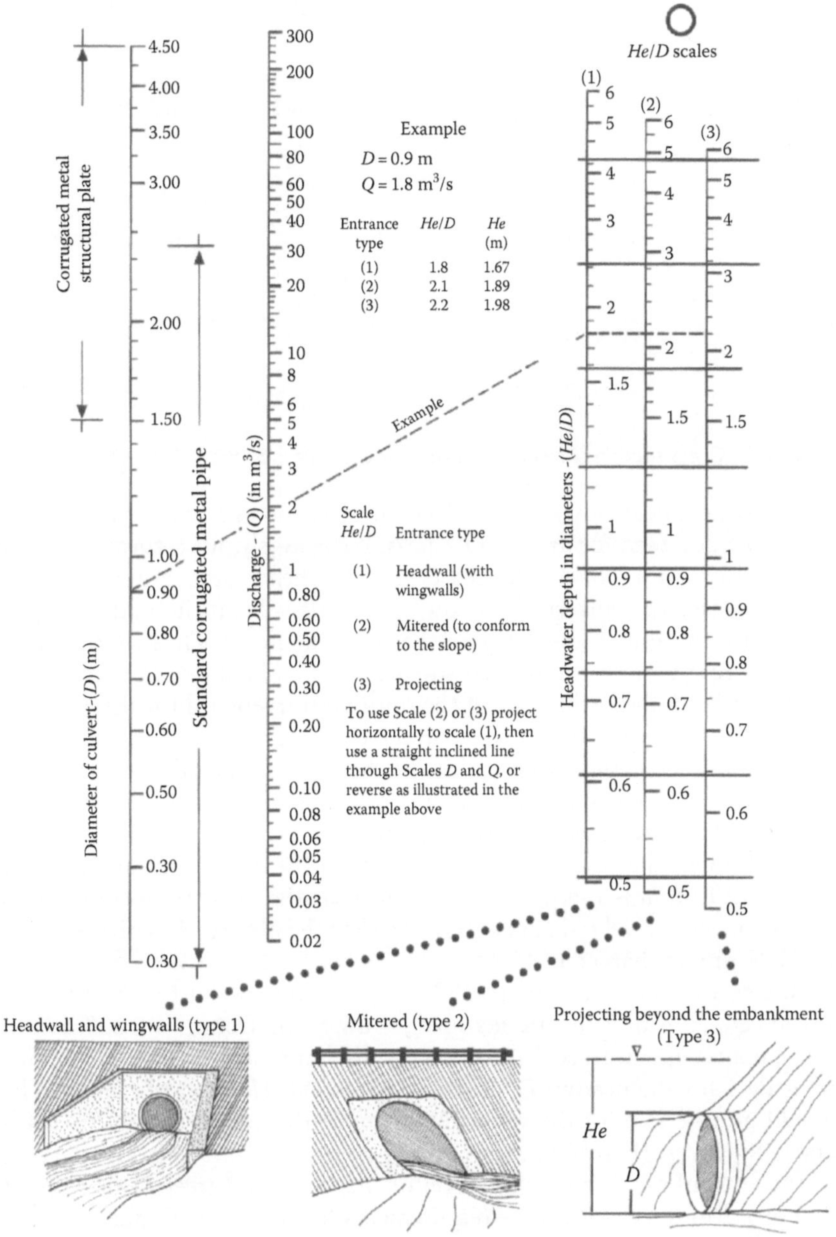

Figure 10.3 Nomograph to find culvert diameter for upstream control case. (From Keller, G. and Sherar, J., *Low-Volume Roads Engineering, Best Management Practices Field Guide*, U.S. Department of Agriculture Forest Service, Plumas National Forest, CA, 158pp., 2003. With the permission of Gordon Keller.)

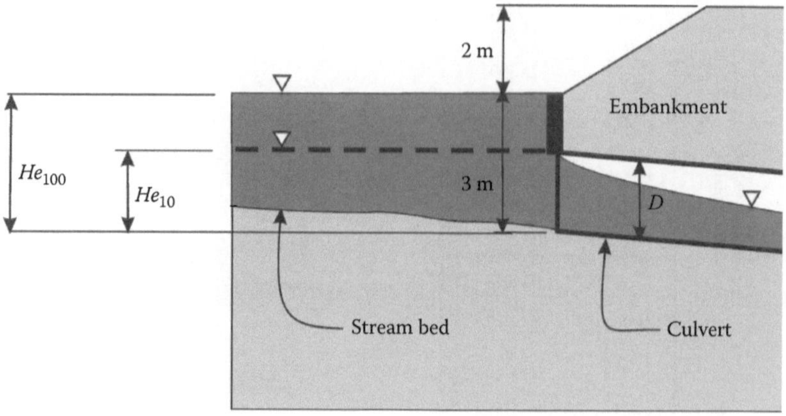

Figure 10.4 Cross section of road and culvert for example in Section 10.4.1. (Not to scale.)

Suppose the peak flows of the culvert shown in cross section in Figure 10.4 should be designed for are 2 and 4 m³/s, respectively, for return periods of 10 and 100 years. For a balanced design, the headwater should rise to be just even with the culvert obvert during the 10-year storm, and below the level of the minimum freeboard for the 100-year storm. Thus, *He*/*D* should be no more than 1 for the 10-year storm and should not be so large as to put the water level above the level of the minimum freeboard for the 100-year storm, 2 m below the road surface in the example.

In Figure 10.3, the three scales on the right-hand side of the nomograph reflect the end conditions shown in the sketches in the middle of the nomograph. Starting with *He*/*D* = 1 on Scale 1 of the nomograph, draw a line through the discharge (*Q*) at 2 m³/s to the diameter scale. A diameter of 1200 mm is found. Therefore, to pass the 10-year storm with the upstream water level rising just to the level of the obvert, the culvert should be 1200 mm in diameter.

To check the adequacy of a 1200 mm diameter culvert for the 100-year storm (Q_{100} = 4 m³/s in this example), draw a line from *D* = 1200 mm on the diameter of culvert scale through the discharge scale at *Q* = 4 m³/s, over to Scale 1 for the headwater depth in diameters, *He*/*D* = 2 is found. This would put the level of the water upstream of the culvert 2 × 1.2 = 2.4 m above the invert (bottom) of the culvert, or 5 − 2.4 = 2.6 m below the crest of the embankment, below the level of the minimum freeboard. A diameter of 1.2 m thus satisfies the requirements for both the 10- and 100-year storms for inlet control.

10.4.2 Culvert installation

Frequently culverts fail not for being the wrong size but because of poor installation. With an understanding of proper installation concepts and techniques, washouts can be avoided. Important features of successful culvert installations are the following:

- Bedding
- Backfill
- Compaction
- Cover

The material for this section has largely been obtained from an excellent book published by the American Iron and Steel Institute [10.4]. Corrugated steel pipe culverts will be the focus, although the principles discussed can be extended to plastic and concrete pipe piles.

10.4.2.1 Bedding

The layer of coarse-grained material on which the culvert pipe sits—the bedding—should be firm yet not so hard as to crush the corrugations in the pipe. Sand, gravel, or crushed rock with a maximum particle size of 19 mm (3/4 in.) make good bedding materials. Compaction should range between 95% and 98% of standard Proctor maximum dry density (SPMDD). The bedding should be at least 200 mm (8 in.) deep or 4% of the fill height over the pipe, whichever is larger. The bedding layer should be 3 pipe diameters wide. Where a pipe spans both rock and soil subgrades, a tapered transition at least 4 pipe diameters long centered on the interface of the rock and soil should be provided.

10.4.2.2 Backfill

The backfill around the pipe is actually what provides the greater part of the strength in the pipe/soil system, not the pipe itself. Properly constructed, the backfill forms a sort of soil arch, carrying the loads on the installation around the pipe and into the ground beside and below the pipe. Therefore, the selection of the backfill material, placement, and compaction are important.

Backfill material should be good clean soil free of organics, peat, frozen lumps, snow, ice, or rocks which will damage the pipe. The backfill should be raised equally on both sides of the pipe. Placement and compaction should be carried out parallel to the pipe.

If the backfill material is especially coarse grained, seepage may occur outside the pipe, in the backfill, raising the risk of the backfill being washed out. Seepage collars formed of clayey material at least 4 pipe diameters long can be built into the backfill at each end, extending down through the bedding and around the pipe at least a pipe diameter wide.

10.4.2.3 Compaction

Because stress concentrates at the lower level of the backfill beside the pipe, particular attention should be paid to the "haunches," the narrow area where the backfill must be placed beside the pipe and the bedding. Compaction can be achieved with hand equipment such as rammers, plate tampers, or "jumping jacks."

It is important to compact the backfill enough to provide the support needed, but not so much as to roll the pipe off its bedding, or distort it or even buckle it. The manufacturer's recommendations for the lower and upper limits on compaction should be followed. For large culverts, limits are placed on the change in the vertical diameter of the pipe, which first increases as backfill is placed and compacted, squeezing the culvert laterally, and then decreases as the placement of the backfill is built over the obvert of the pipe. The change in pipe diameter can be monitored by hanging a plumb bob from the obvert, with the plumb bob hanging close to the invert of the pipe. The distance from the plumb bob to the invert is measured periodically as back-filling proceeds, and the change in diameter checked against the manufacturer's limits.

10.4.2.4 Cover

Too often not enough cover is provided over the culvert obvert, the soil arching effect cannot develop, and the installation is crushed by traffic. This blocks the watercourse, and the culvert is eventually overtopped or blown out, with consequent damage to aquatic habitat and organisms. Adequate cover is *D*/4 or 600 mm (2 ft), whichever is larger.* This should provide enough strength in the pipe/soil system to carry conventional axle loads and for gravel roads, enough additional depth of gravel to account for the loss due to grading maintenance. If axle loads are to be large, a specific analysis of the load-carrying capacity of the installation should be performed.

10.5 BRIDGES

Beam bridges, arches, suspension bridges, and cable-stayed bridges can be used. Beam bridges can be designed as single spans, or multiple spans simply supported or continuous over one or more piers [10.2] (Figure 10.5). If foundation conditions are nonuniform, simply supported spans are

* These limits are double those usually quoted for culverts in conventional roads [10.4]. The reasons to double the cover requirement are (1) that LVR are often have unsealed surfaces which are graded periodically, potentially resulting in a loss of cover, and (2) LVR can be subjected to much heavier loads than their conventional counterparts, requiring a stronger soil "arch" over them.

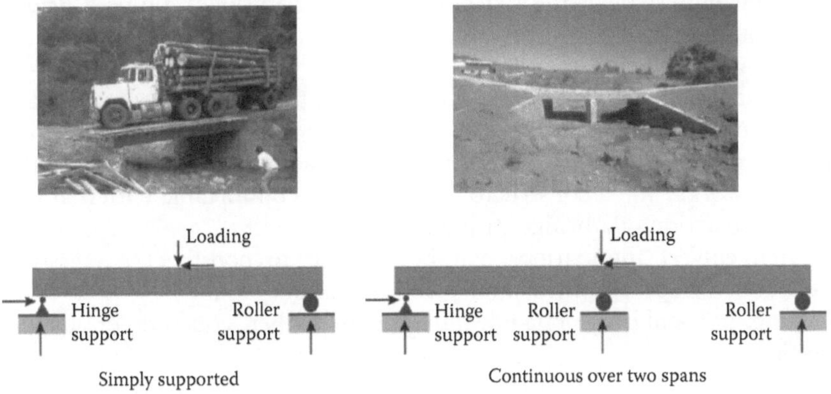

Figure 10.5 Beam bridge types. Arrows indicate loads and reactions. The continuous span bridge can be damaged by settlement of one of the supports. (Photos used with permission of Gordon Keller.)

favored, as they are less sensitive to differential settlement than multispan continuous bridges. Bridge configurations that require in-water piers can be problematic, causing jams of debris or ice.

Often bridges that fail for other than structural reasons fail because of scour, or because they have an inadequate opening size for the peak flow they are to pass [10.1]. They should be designed with enough capacity to pass a 100- or 200-year peak flow. Bridges are expensive compared to other watercourse crossing structures, and the consequences of their failure can be great, given the importance to the road network they serve, so an approach to their design more conservative than that used for fords and culverts is warranted.

10.5.1 Advantages and disadvantages

Why specify a comparatively expensive bridge rather than some less expensive alternative? From an aquatic habitat point of view, bridges are generally better than the alternatives; construction can take place outside the watercourse bottom, and the bottom is the same as the rest of the watercourse, unaltered. However, bridges are generally more expensive than the alternatives.

Simple bridges can be erected fairly quickly, and prefabricated bridges are available in standard kits. They must be suited to the foundation requirements, hydraulic requirements, and traffic conditions (vehicle size, weight, axle loads), however. Concrete bridges are desirable for their long life (100+ years) with little maintenance [10.1]. However, the logistics of constructing a concrete bridge on site may be challenging—the construction site must be less than about 2 h travel time from the concrete plant or the concrete will have begun to set before it reaches the construction site.

For medium spans, large arch culverts and short bridges compete. Bridges can require greater technical expertise for the constructor than large culverts, unless the culvert is a very large bolted multiplate unit. Compared to the abutments and approaches to a bridge, the embankment for a large culvert may require a much greater amount of fill, which may be difficult to find on site. On the other hand, culvert installations are primarily a matter of earthworks, and a constructor may be more comfortable with that than the construction of a bridge structure.

Arch culvert foundations can be difficult to construct successfully. However, bridge abutment and pier foundations might be equally challenging, particularly if piled foundations of timber, concrete, or steel are needed.

10.5.2 Bridges: Good practice

Bridge design should be accomplished by a team with a geotechnical engineer, a structural engineer, a hydrologist, and a fisheries biologist. A thorough study of the candidate locations should be made, from the points of view of the hydrology, traffic, and biota. The design should satisfy all of these disciplines.

Bridge foundations are critical. The most appropriate selection between spread footings or piling should be made. If piled foundations are called for, consideration should be given to driven timber, steel or concrete piles, or bored concrete piles. Whatever foundation system is selected, it must be protected from scour (Figure 10.6). Concrete wing walls, gabion baskets,

Figure 10.6 Scour of a bridge foundation. The right abutment was completely scoured away. The bridge has been founded on temporary piers, while a new abutment is constructed.

Figure 10.7 Poorly designed bridge: the bridge is not well aligned with the stream, and it constricts stream flow.

and/or rip rap should be deployed to protect critical parts of the bridge's foundations. The span should be long enough not to constrict the flow of the watercourse (Figure 10.7). Freeboard of at least 0.5–1.0 m (2–3 ft) beneath the beams of the bridge should be provided to permit the passage of debris or ice during high flows.

Bridge assets must be maintained. Inspections should be carried out by competent engineers on a cycle of 2 years or less, depending on the characteristics of the traffic (gross weight, axle loads, number of passes per unit of time).

10.6 CROSSING WETLANDS

A special case arises when crossing wetlands, muskeg (peatlands), or meadows. These broad flat areas, with ill-defined drainage, present special challenges. The subgrades can be soft, weak, and deep, and it can be unclear where to place the needed water crossing. Sometimes it isn't clear which way the water flows, or it may even reverse direction from time to time. Wetlands, muskeg, and meadows can be rich in aquatic organisms, and roads through them can disrupt the habitat and movement of the organisms.

The best solution to the problem is to avoid the area in the first place, seeking a better route. The route around will be longer, but over the life of the road may be less expensive. Building on wetlands, muskeg, or meadows often requires deep embankments which settle metres (yards) into the soft subgrade until an equilibrium is reached. Any culvert in the embankment must be designed to settle metres (yards) too.

If for some reason a commitment is made to crossing the deposit, it may be that building the required embankment from a porous material such as open-graded crushed rock will provide enough cross flow through the embankment. Assuming the subgrade is a soft, fine-grained material, the embankment fill will have to be filtered, so it does not become clogged with the fines of the subgrade. See Section 8.3.8 for filter design.

If the subgrade is firm and can support a structure, numerous smaller-diameter culverts can be installed. These should be spread out along the watercourse crossing, to make a low embankment possible, to make compaction between culverts effective, and to accommodate the broad drainage area without unnaturally concentrating the flow. A vented ford could be a good solution. However, if significant amounts of debris coming through the watercourse during storm events, the multiple small-diameter culverts will need to be cleared promptly.

If bedrock is within a shallow depth, it may be possible to construct a ford, building up from a rock foundation.

10.7 WATERCOURSE CROSSING LOCATION

A good watercourse crossing has the following characteristics:

- It crosses the watercourse at as close to 90° as possible
- It crosses the watercourse where it is relatively straight
- It is located where the watercourse has stable banks, where there is little or no erosion
- It follows the original channel of the watercourse, rather than requiring the training of the watercourse into a different alignment
- It doesn't constrict the flow (Figure 10.7)
- It has an open bottom so the watercourse bottom is unchanged within the limits of the crossing, maintaining its habitat and conditions
- It is located where there is little or no debris or ice to contend with
- Its road approaches are aligned well for traffic, with gentle curves, long sight lines, and flat grades

Having the watercourse and road cross at approximately 90° results in a shorter ford, culvert, or bridge. If the watercourse is straight and the banks stable, the flow conditions in the channel are likely to be stable, and unlikely to undercut bridge abutments or culvert embankments. Following the original channel will avoid diverting the flow, reducing the risk of undercutting abutments or approach fills. Constricting the flow through the bridge or culvert, or over the ford will increase the flow velocity in the watercourse, increasing the erosive power of the water.

Designing the watercourse crossing with an open bottom will provide friendlier conditions to aquatic organisms. Ice and debris are to be avoided as they can jam the culvert or bridge opening and back up the water, putting added pressure on abutments or fills, potentially overtopping or blowing out the bridge or culvert.

Finally, the watercourse crossing must be designed with vehicle traffic in mind. Because watercourse crossings are necessarily narrower than the rest of the road, they are inherently riskier than the rest of the road. Sudden changes in grade or horizontal alignment, or poor sight lines add to the risk.

REFERENCES

10.1 Keller, G. and Sherar, J. 2003. *Low-Volume Roads Engineering, Best Management Practices Field Guide*. Plumas National Forest, CA: U.S. Department of Agriculture Forest Service, 158pp.

10.2 Lay, M.G. 2009. *Handbook of Road Technology*, 4th edn. Abingdon, U.K.: Spon Press, 933pp.

10.3 O'Flaherty, C.A. (ed.). 2002. *Highways, the Location, Design, Construction, and Maintenance of Pavements*. Oxford, U.K.: Butterworth-Heinemann, 553pp.

10.4 AISI. 1984. *Handbook of Steel Drainage and Highway Construction Products*, Canadian edition. Washington, DC: American Iron and Steel Institute, 413pp.

Chapter 11

Slope stability

11.1 INTRODUCTION

11.1.1 Assumptions

It is assumed in this chapter that the reader fully understands basic soil mechanics including cohesive and cohesionless soils, effective and total stress analysis, seepage, and flow nets. If not, the reader is directed to basic soil mechanics texts such as Craig [11.1].

11.1.2 Types of slopes, causes of failures, types of slope failures

Significant slopes—those over 4.6 m (15 ft) in height—occur naturally in road work, or may be the result of cutting or filling operations. Angles for slopes less than this height are usually simply fixed at a figure such as 2:1 or 3:1 (horizontal:vertical) or some other value based on experience with the soil of which they are constructed; however, angles for slopes greater than this height are usually determined through analysis and design. Their design is based on the premise that gravitational and seepage forces may cause instability. Superimposed loads such as those of equipment, buildings, roads, vehicles, railways, canals, tailings piles, or stockpiles may also add to the instability and must be included in the analysis supporting the design.

Slope failure modes can be classified as plane translational, rotational, or compound [11.1] (Figure 11.1). Plane translational failures are typically shallow and parallel to the ground surface. They are often controlled by a stronger stratum at shallow depth or by preexisting slips or fissures. The failure surface in rotational failures is at greater depth and may be circular or noncircular. Circular failures are associated with homogeneous, isotropic soil conditions; noncircular failures are associated with nonhomogeneous, anisotropic conditions [11.1]. Compound failures are often controlled by stronger strata at great depth.

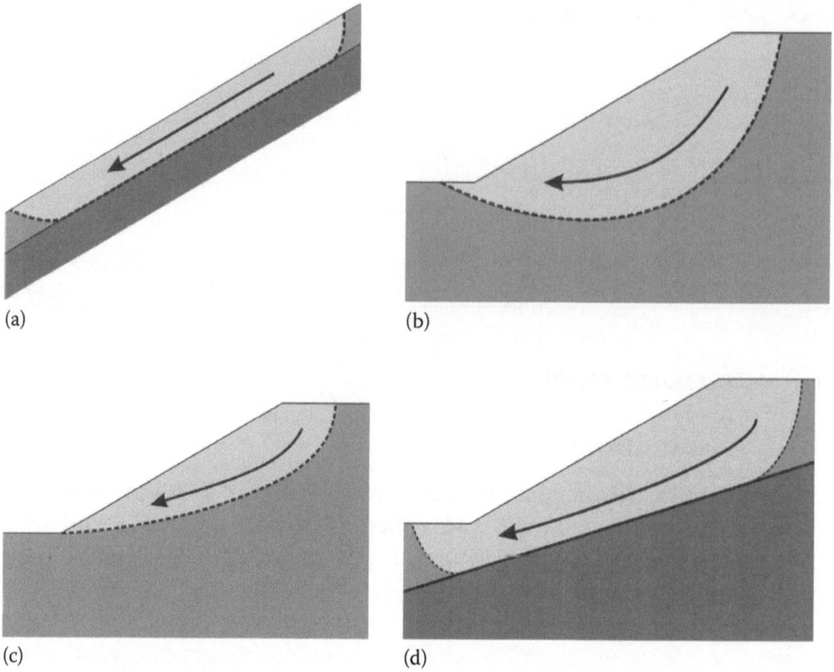

Figure 11.1 Types of slope instability: (a) plane translational failure, (b) circular rotational failure, (c) noncircular rotational failure, (d) compound failure.

11.1.3 Analyses

In many cases, the analysis is two-dimensional, in plane strain. Limit equilibrium is applied to a known or assumed surface about to fail. The shear strength available is compared to the shear resistance needed for stability, and an overall factor of safety is determined:

$$F = \frac{\tau_f}{\tau} \quad (11.1)$$

where
- F is the overall factor of safety (unitless)
- τ_f is the shear resistance available (kN/m^2 or lb/ft^2)
- τ is the shear resistance needed for the slope to remain stable (kN/m^2 or lb/ft^2)

The failure surface with the lowest factor of safety is determined by iteration, with the analysis repeated for dozens or even hundreds of potential failure surfaces. It is now rare to make the analysis by hand; designers usually turn to software with computerized versions of the methods to be described below.

Shear is not the only condition to be checked. Slopes may also fail by surface erosion, internal erosion due to high seepage gradients or poor compaction, excessive deformation, or bearing capacity failure in the underlying soil [11.1]. These are not covered here.

11.2 PLANE TRANSLATIONAL FAILURES

Plane translational failures are modeled as shallow failures in "infinite" slopes (Figure 11.2). The derivation of a stability analysis for them initially assumes $\phi - c$ soil and a flow net for seepage parallel to the ground surface [11.1]. Equation 11.1 defines a factor of safety. The shear strength available is defined in Equation 11.2:

$$\tau_f = (\sigma - u)\tan\phi' + c' \tag{11.2}$$

where

- τ_f is shear resistance available (kN/m^2 or lb/ft^2)
- σ is the normal stress (kN/m^2 or lb/ft^2)
- u is the porewater pressure (kN/m^2 or lb/ft^2)
- ϕ' is the effective angle of shearing resistance
- c' is the effective cohesion (kN/m^2 or lb/ft^2)

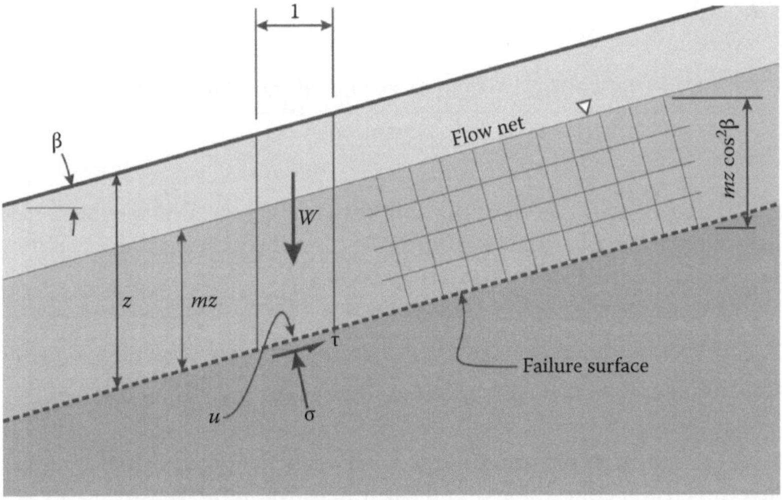

Figure 11.2 Plane translational failure.

The contributing equations for σ, u, and τ are given in Equations 11.3 through 11.5 [11.1]:

$$\sigma = \{(1-m)\gamma + m\gamma_{sat}\} z\cos^2\beta \quad (11.3)$$

$$u = mz\gamma_w \cos^2\beta \quad (11.4)$$

$$\tau = \{(1-m)\gamma + m\gamma_{sat}\} z\sin\beta\cos\beta \quad (11.5)$$

where

σ is the normal stress (kN/m^2 or lb/ft^2)
u is the porewater pressure (kN/m^2 or lb/ft^2)
τ is the shear resistance needed for the slope to remain stable (kN/m^2 or lb/ft^2)
m is defined in Figure 11.2
z is the vertical depth to the failure surface (m or ft)
γ is the moist unit weight of the soil (kN/m^3 or lb/ft^3)
γ_{sat} is the saturated unit weight of the soil (kN/m^3 or lb/ft^3)
γ_w is the unit weight of water (9.81 kN/m^3 or 62.4 lb/ft^3)
β is the angle between the ground surface and horizontal

Substituting into Equation 11.1, the expression for the factor of safety against plane translational failure becomes

$$F = \frac{\tau_f}{\tau} = \frac{(\sigma - u)\tan\phi' + c'}{\{(1-m)\gamma + m\gamma_{sat}\} z\sin\beta\cos\beta}$$
$$= \frac{\left[\{(1-m)\gamma + m\gamma_{sat}\} z\cos^2\beta - mz\gamma_w\cos^2\beta\right]\tan\phi' + c'}{\{(1-m)\gamma + m\gamma_{sat}\} z\sin\beta\cos\beta} \quad (11.6)$$

If $c' = 0$ and $m = 0$, meaning the soil between the ground surface and the failure surface is cohesionless and moist (the phreatic surface is below the failure surface), the factor of safety becomes

$$F = \frac{\tan\phi'}{\tan\beta} \quad (11.7)$$

If $c' = 0$ and $m = 1$, meaning the soil between the ground surface and the failure surface is cohesionless and saturated (the phreatic surface is at the ground surface), the factor of safety becomes

$$F = \frac{\gamma' \tan\phi'}{\gamma_{sat} \tan\beta} \tag{11.8}$$

Note that in Equations 11.7 and 11.8, where in both cases $c' = 0$, the factor of safety against failure is independent of the depth z. If $c' \neq 0$, F is a function of z, and β can be greater than ϕ' if z is less than some critical value.

In a total stress (undrained) analysis, c' and ϕ' are replaced by c_u and ϕ_u, and u is set to zero.

11.3 CIRCULAR FAILURES, $\phi_U = 0$

The case for a slope in fully saturated cohesive soil immediately after construction is dealt with using an undrained, total stress analysis. A circular failure surface with center O, radius r, and arc length L_a is assumed (Figure 11.3) and moment equilibrium at the instant of shear failure is considered. In Equation 11.1, the shear resistance available is the undrained cohesion:

$$F = \frac{c_u}{\tau} \tag{11.9}$$

where

F is the overall factor of safety (unitless)

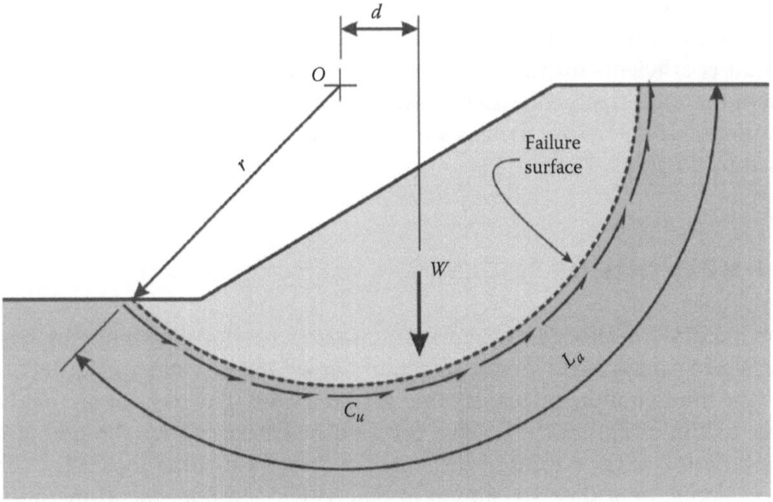

Figure 11.3 Circular rotational failure in fully saturated cohesive soil.

c_u is the shear resistance available (kN/m² or lb/ft²)
τ is the shear resistance needed for the slope to remain stable (kN/m² or lb/ft²)

The weight of the sliding mass is *W*, acting at a distance *d* from the center of the failure surface O. Equating moments about the center:

$$F = \frac{c_u L_a r}{W d} \tag{11.10}$$

where
F is the overall factor of safety (unitless)
c_u is the undrained cohesion (kN/m² or lb/ft²)
L_a is the length of the failure surface (m or ft)
r is the radius of the failure surface arc (m or ft)
W is the weight of the sliding mass (kN/m or lb/ft)
d is the distance from the line of action of *W* from the center of the failure surface arc O

If a tension crack opens at the top of the failure surface and fills with water, the failure arc L_a is shortened by the depth of the crack, taken to be 2 c_u/γ, and the hydrostatic force applied [11.1]. In practice, a large number of combinations of failure arc radii and center locations are assessed, searching for the minimum factor of safety. Usually a factor of safety of at least 1.3 is sought.

Taylor [11.2] developed stability coefficients N_s from which the factor of safety can be determined for slope stability in saturated cohesive soils overlying a hard stratum [11.1]. Gibson and Morgenstern [11.3] published similar coefficients for the case where normally consolidated cohesive soils where the undrained cohesion c_u varies linearly with depth. Design charts accommodating factors such as surcharges and slope submergence are provided by TRB [11.4].

11.4 METHODS OF SLICES

The method of slices is a general approach to slope stability analysis. The sliding mass is divided into a number of vertical slices (Figure 11.4) and the limit equilibrium considered for each slice. The basis of the method is the assumption that the normal force at the base of a slice should mainly be a function of the weight of the soil above the base of the slice [11.5]. The forces acting on a slice are shown in Figure 11.5. In the general method of slices, the equilibrium of each slide is considered, to determine the normal effective force and shear force acting on the base of the slice. The overall

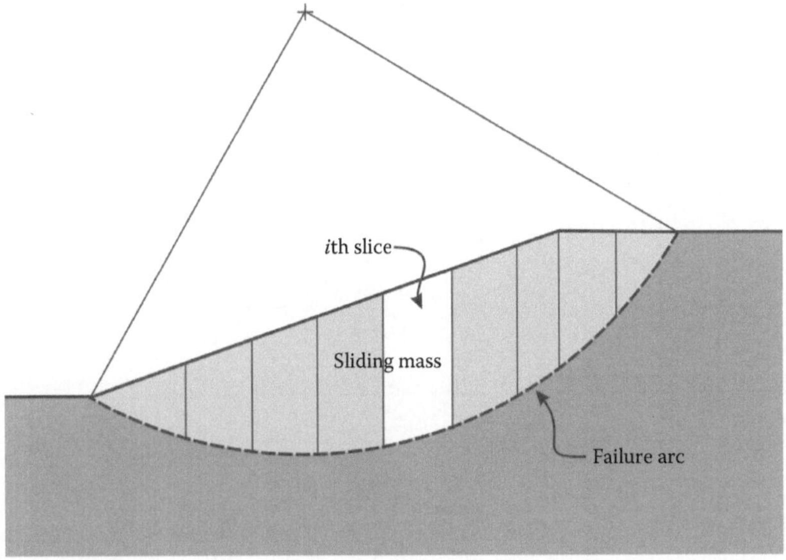

Figure 11.4 Slope failing on circular arc divided into slices (*i*th slice shown in Figure 11.5).

factor of safety is then defined in terms of moments about the center of the failure arc [11.5]

$$F = \frac{\text{moment of shear strength along failure arc}}{\text{moment of weight of failing mass of soil}} \tag{11.11}$$

or

$$F = \frac{\tan\phi' \sum_{i=1}^{n} N_i' + c'L}{\sum_{i=i}^{n} W_i \sin\theta_i} \tag{11.12}$$

where

- F is the factor of safety (unitless)
- ϕ' is the effective (drained) angle of friction
- N_i is the normal effective force on the base of the slice (kN/m or lb/ft)
- c' is the effective (drained) cohesion (kN/m^2 or lb/ft^2)
- L is the length of the failure arc (m or ft)
- W_i is the weight of the slice (kN/m or lb/ft)
- θ_i is the angle between the base of the slice and horizontal

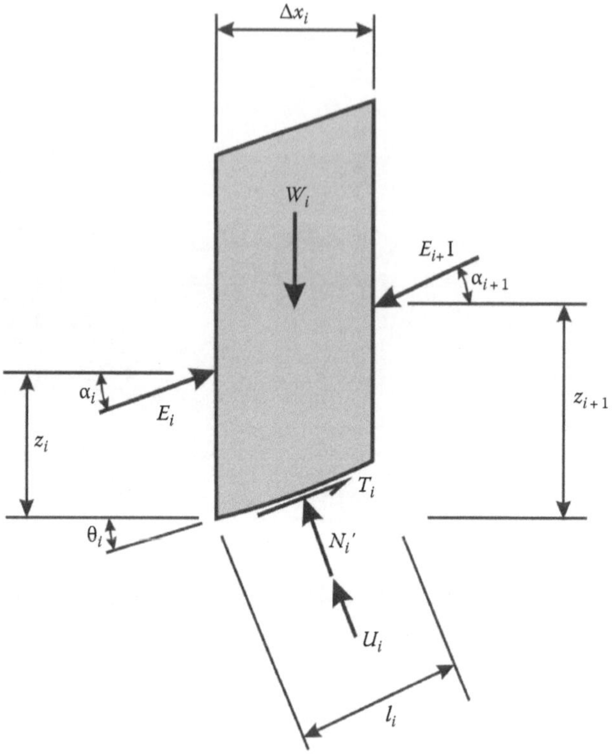

Figure 11.5 Forces on *i*th slice: W_i is the weight of the slice, E_i *and* $E_{i+}1$ are the forces imposed by adjacent slices, N'_i *and* U_i are the normal effective force and porewater force, respectively, acting on the base of the slice (porewater pressure determined from flownet, not shown), and T_i is the shear force acting on the base of the slice.

The problem is statically indeterminant, with $5n - 2$ unknowns, but only $3n$ available equations; $2n$ assumptions must be made. The various methods are distinguished by the assumptions they make, which in turn affects their applicability and accuracy. Table 11.1 [11.4] lists some of the commonly used solutions.

11.5 BISHOP'S MODIFIED METHOD OF SLICES

The method will serve as an example of the implementation of a method of slices. Bishop's modified method of slices assumes that the reactions between slices are horizontal, that is, α_i and α_{i+1} are 0 in Figure 11.5. Note also that plane strain is assumed, with the slice being one unit (1 m or 1 ft) thick.

Table 11.1 Selected methods of slices for slope stability

Method	*Year*	*Applicability*
Ordinary method (Fellinius method)	1927	Circular failure surfaces only
Bishop's modified method	1955	Circular failure surfaces only
Lowe and Karafiath's method	1960	Circular and noncircular failure surfaces
Morgenstern and Price's method	1965	Circular and noncircular failure surfaces
Spencer's method	1967	Circular and noncircular failure surfaces
Janbu's simplified method	1968	Circular and noncircular failure surfaces
Janbu's generalized method	1968	Circular and noncircular failure surfaces
Modified Swedish method	1970	Circular and noncircular failure surfaces
Sarma's method	1973	Circular and noncircular failure surfaces

Source: TRB, *Landslides—Investigation and Mitigation*, Turner, A.K., and Schuster, R.L. (eds.), U.S. Transportation Research Board Special Report 247, Transportation Research Board, U.S. National Academy of Sciences, Washington, DC, 673pp., 1996.

With these assumptions, the equation for the equilibrium of the forces acting in the vertical direction on a slice can be written [11.5]

$$N_i' = \frac{W_i - u_i\,\Delta x_i - (c'/F)\Delta x_i \tan\theta_i}{\cos\theta_i\left[1 + \tan\theta_i\ (\tan\phi_i'/F)\right]} \tag{11.13}$$

where

N_i' is the normal effective force acting on the base of the slice (kN/m or lb/ft)
W_i is the weight of the slice (kN/m or lb/ft)
u_i is the porewater pressure acting on the base of the slice (kN/m^2 or lb/ft^2)
Δx_i is the width of the slice (m or ft)
c' is the effective (drained) cohesion (kN/m^2 or lb/ft^2)
F is the factor of safety (unitless)
θ_i is the angle between the base of the slice and horizontal
ϕ' is the effective (drained) angle of shearing resistance

Substituting for N_i' in Equation 11.12 the overall factor of safety in terms of moments becomes

$$F_{calculated} = \frac{\sum_{i=1}^{n}\left[c'\Delta x_i + \left(W_i - u_i\,\Delta x_i\right)\ \tan\phi'\right]\left[{}^{1}\!/_{M(\theta_i)}\right]}{\sum_{i=1}^{n} W_i \sin\theta_i} \tag{11.14}$$

with

$$M(\theta_i) = \cos\theta_i \left(1 + \tan\theta_i \frac{\tan\phi'}{F_{assumed}}\right) \tag{11.15}$$

where

$F_{calculated}$ is the calculated factor of safety (unitless)
$F_{assumed}$ is the factor of safety first assumed (unitless)
c' is the effective (drained) cohesion (kN/m² or lb/ft²)
Δx_i is the width of the slice (m or ft)
W_i is the weight of the slice (kN/m or lb/ft)
u_i is the porewater pressure acting on the base of the slice (kN/m² or lb/ft²)
ϕ' is the effective (drained) angle of shearing resistance
θ_i is the angle between the base of the slice and horizontal

Note that the factor of safety is in fact a function of the factor of safety. To solve the problem, an iterative approach is taken, where a factor of safety is assumed, $M(\theta_i)$ calculated (Equation 11.15) and substituted into Equation 11.14, which is solved for F. The calculated F is compared to the assumed F, a new F assumed, and the process is repeated until there is convergence.

Example calculations are shown in Tables 11.2 and 11.3. Both tables apply to Figure 11.4, with slope height, failure arc radius, and soil unit weight equal to 6 m (20 ft), 15.9 m (52 ft), and 20 kN/m³ (130 lb/ft³), respectively. In Table 11.2, a simple case with c_u = 30 kN/m² (625 lb/ft²) and $\phi_u = 0$ is examined, and a factor of safety against rotational failure $F = 2.05$ calculated. Note that iteration is not required if $\phi = 0$. If so, $M(\theta)$ reduces to just cos θ in Equation 11.15, independent of the assumed factor of safety, $F_{assumed}$. The same case can be calculated using Equation 11.10. Using this approach, the calculated factor of safety is 1.97 (a difference of 4%).

In Table 11.3, the factor of safety for the same slope and assumed failure surface, for c' = 1 kN/m² (21 lb/ft²) and ϕ' = 20°, is calculated to be 1.84 ($F \geq 1.3$ is a common criterion for acceptance). Convergence occurs with 2 iterations.

11.6 SOFTWARE

Now, it is rare that a designer would carry out a manual analysis, except perhaps for a "back of the envelope" calculation. Inherent in any design is the necessity to analyze many—as in dozens or hundreds—potential failure surfaces or arcs, with the necessity of finding the one with the lowest factor of safety. The designer turns to commercial software.

Table 11.2 Example problem, slope stability analysis using Bishop's modified method: $c_u = 30\ kN/m^2$, $\phi_u = 0°$

								Assumed $F = 2.00$	
Slice	Δx_i *(m)*	θ_i	W_i *(kN/m)*	u_i *(kN/m²)*	c_u *(kN/m²)*	ϕ_u	$W_i \sin \theta_i$ *(kN/m)*	$M(\phi_i)$	*Numerator (kN/m)*
1	3.5	−24	95	0	30	0	−39	0.914	115
2	2.4	−11	160	0	30	0	−31	0.982	73
3	2.4	−3	220	0	30	0	−12	0.999	72
4	2.4	5	250	0	30	0	22	0.996	72
5	2.4	13	270	0	30	0	61	0.974	74
6	2.4	23	270	0	30	0	105	0.921	78
7	1.5	32	160	0	30	0	85	0.848	53
8	2.1	40	170	0	30	0	109	0.766	82
9	2.3	50	70	0	30	0	54	0.643	107
							$\Sigma = 355$		$\Sigma = 727$
								Calculated $F = 2.05$	

Table 11.3 Example problem, slope stability analysis using Bishop's modified method: $c' = 1$ kN/m², $\phi' = 20°$

								Assumed $F = 1.00$		Assumed $F = 1.78$		Assumed $F = 1.84$	
Slice	Δx_i (m)	θ_i	W_i (kN/m)	u_i (kN/m²)	c'_i (kN/m²)	ϕ'	$W_i \sin \theta_i$ (kN/m)	$M(\theta_i)$	Numerator (kN/m)	$M(\theta_i)$	Numerator (kN/m)	$M(\theta_i)$	Numerator (kN/m)
1	3.5	−24	95	0	1	20	−39	0.766	50	0.830	46	0.833	46
2	2.4	−11	160	0	1	20	−31	0.912	66	0.943	64	0.944	64
3	2.4	−3	220	0	1	20	−12	0.980	84	0.988	83	0.988	83
4	2.4	5	250	0	1	20	22	1.028	91	1.014	92	1.013	92
5	2.4	13	270	0	1	20	61	1.056	95	1.020	99	1.019	99
6	2.4	23	270	0	1	20	105	1.063	95	1.000	101	0.998	101
7	1.5	32	160	0	1	20	85	1.041	57	0.956	62	0.953	63
8	2.1	40	170	0	1	20	109	1.000	64	0.897	71	0.893	72
9	2.3	50	70	0	1	20	54	0.922	30	0.799	35	0.794	35
							Σ = 355		Σ = 633		Σ = 654		Σ = 655
								Calculated F = 1.78		**Calculated F = 1.84**		**Calculated F = 1.84**	

The software itself is just a computerized form of one of the methods such as those listed in Table 11.1. The interfaces are convenient to varying degrees from one package to another, and the analysis methods themselves have various strengths and weaknesses. See TRB [11.4] for a full discussion.

11.7 SELECTION OF METHOD

Undrained and drained (total and effective stress) analyses have been discussed earlier. Guidance is needed in the selection of the analysis type.

Stresses and porewater pressures in soil slopes change with changes in external loading. External loads change with the rapid construction of embankment slopes, with the rapid excavation of cut slopes, with the inundation of slopes or the rapid drawdown against slopes. The speed of the change is relative to the permeability of the soil of which the slope is constructed.

If the stability under undrained conditions, immediately after the construction of an embankment slope or the excavation of a cut slope, is required, it is assumed that the undrained shear strength exhibited in tests of small specimens in the laboratory will be the same for the same soil in the slope subjected to the same conditions, and that the undrained laboratory shear strengths can be used in the slope stability analysis. The challenge is ensuring that the conditions—including the degree of saturation, density, stress history of the soil, rate of loading, and magnitudes and orientations of applied stresses—are the same [11.4].

If the stability of the slope under drained conditions, long after the external loading has changed and excess porewater pressures have dissipated, is required, an effective stress analysis is applied using the effective strength parameters c' and ϕ' [11.4].

If internal porewater pressures change naturally, for instance, after a long wet period, a drained analysis can be applied using the known porewater pressures [11.4].

11.8 REMEDIAL ACTION

11.8.1 Traditional solutions

Often poor drainage is a source of instability. Regional flow of surface water can enter the slope from above, adding to the weight of the sliding mass and increasing the porewater pressures, weakening the slope. Cut slopes may intercept groundwater flows which break out at or near the surface of the slope. In either case reducing porewater pressures can improve slope stability. Surface water can be intercepted with ditches across the top of the slope, parallel to and set back from the crest (Figure 11.6). These ditches should

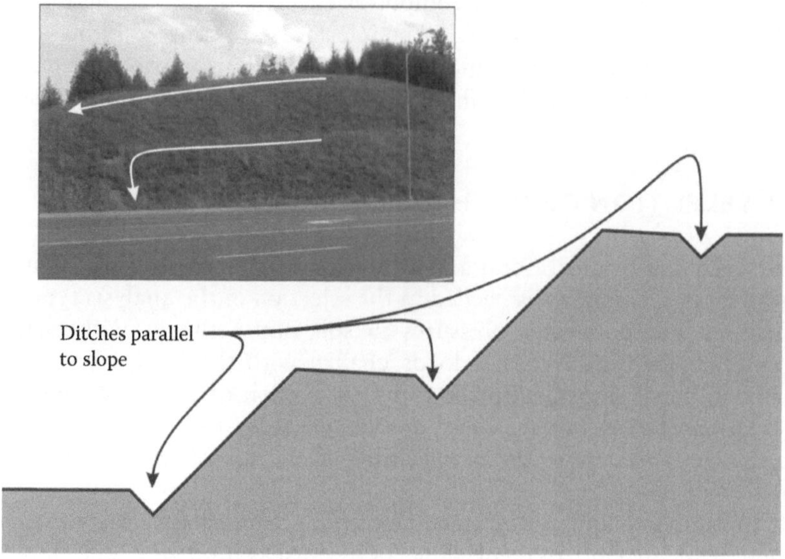

Figure 11.6 Crest and intermediate ditches parallel to the slope divert surface water. Ditch outlets should be properly protected against erosion.

Figure 11.7 Coarse crushed rock ("rip rap") can be used to blanket slopes where groundwater flow breaks out at the surface, causing erosion.

outlet in safe locations and should be protected from erosion. Phreatic surfaces can be lowered and internal porewater pressures reduced by installing internal drains, perhaps accomplished with horizontal drilling. Springs can be blanketed with coarse crushed rock (riprap) (Figure 11.7).

Unstable steep slopes can simply be flattened if space is available. The whole slope may be flattened, or it may be built as a series of benches which reduces the effective angle of the slope (Figure 11.8). The benches provide the opportunity to install intermediate parallel drainage ditches.

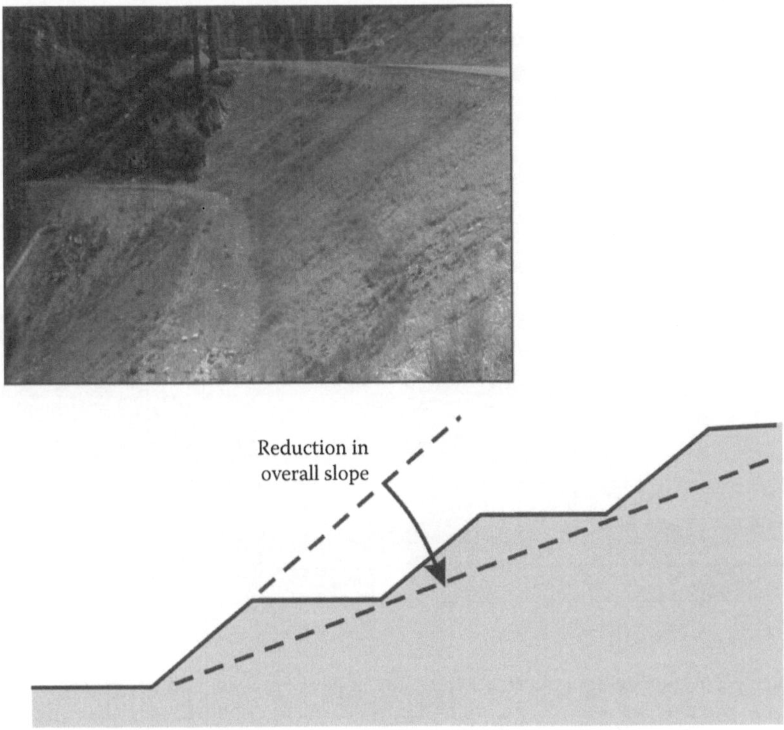

Figure 11.8 Slopes with benches have flatter overall slope angles. (Photo courtesy Gordon Keller, used with permission.)

If space is not available, various retaining structures can be built at the toe of the slope, to accomplish the flattening of the remainder of the slope. The retaining structures commonly used include gravity retaining walls, cantilever retaining walls, gabion basket walls, and crib walls (Figure 11.9).

"Burrito walls"* can be constructed using geosynthetics (Figure 11.10). A granular drainage/foundation layer is first prepared at the bottom of the slope, and then a layer of geosynthetic material placed on the foundation. Next, a layer of free draining fill is placed and compacted on the geosynthetic layer. Then the free end of the geosynthetic is folded over the fill, a second geosynthetic layer placed, and a new layer of compacted fill placed on the second geosynthetic layer. The free end of the second geosynthetic layer is folded over the fill and the process repeated layer by layer until the design height is reached. The walls have the advantage of being tolerant of settlement and differential settlement. The exposed geosynthetic must be covered to reduce the damage due to exposure to ultraviolet rays.

* I believe it was Gordon Keller, retired U.S. Forest Service engineer, who first coined this very apt term—Author.

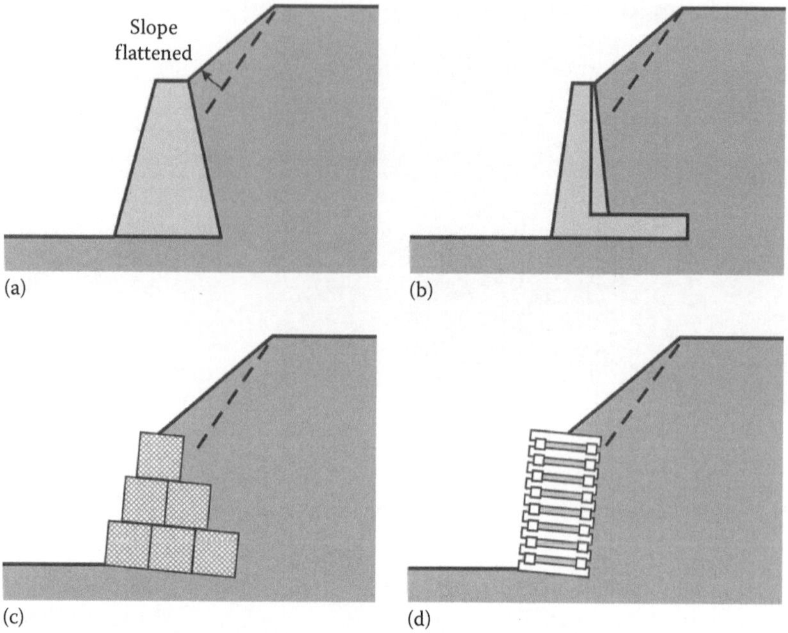

Figure 11.9 Retaining structures such as (a) gravity retaining walls, (b) cantilevered retaining walls, (c) gabion basket walls, or (d) crib walls of timber or precast concrete can be used to flatten slopes.

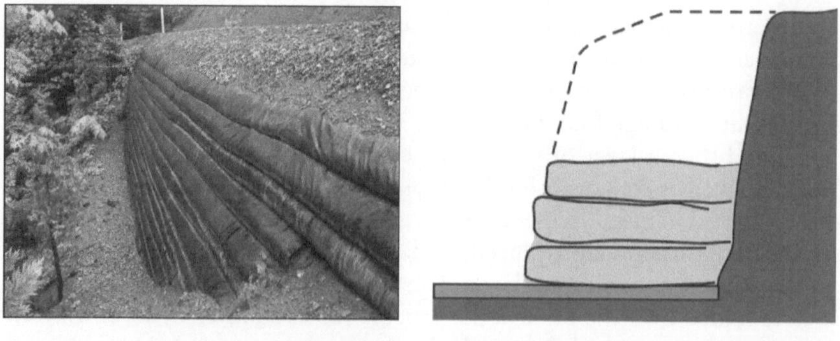

Figure 11.10 "Burrito walls" are constructed, layer by layer, by placing a geosynthetic layer, placing and compacting free-draining fill on the geosynthetic, and then lapping the free end of the geosynthetic over the compacted fill. The process is repeated until the wall reaches design height. (Photo courtesy Gordon Keller, used with permission.)

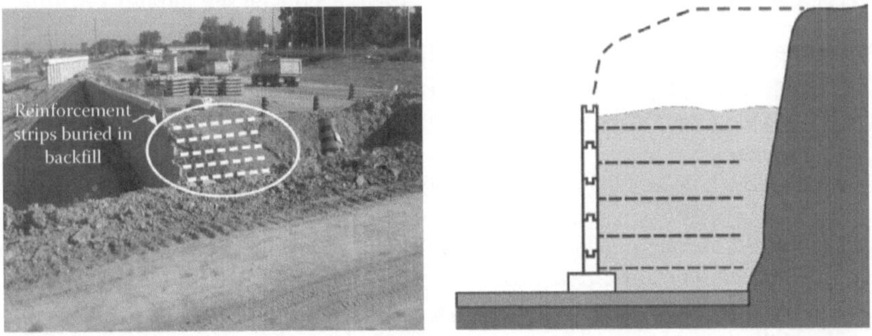

Figure 11.11 Mechanically stabilized retaining walls are constructed from precast concrete facing panels anchored with reinforcing strips or geosynthetic layers extending into the backfill behind the wall.

11.8.2 Reinforcement

Mechanically stabilized earth retaining structures can be constructed (Figure 11.11). These consist of stacked tongue-and-groove precast concrete facing units held in place with metallic or geosynthetic reinforcement elements extending back into the backfill behind the wall.

Other slope reinforcement schemes relying on geosynthetics are discussed at length in Chapter 12.

11.8.3 Bioengineered and biotechnical solutions

Relatively recently slope stabilization schemes which rely on vegetation have been devised. These have the advantages of using local labor and natural materials to provide context-sensitive solutions. A comprehensive treatment of them is given in Fay et al. [11.6].

11.9 OTHER TYPES OF SLOPE FAILURES

Slope failures such as spreads, flows, and avalanches have not been covered here. The reader is directed to TRB [11.4] and Rodriguez et al. [11.7] for a complete treatment of them.

REFERENCES

11.1 Craig, R.F. 1997. *Soil Mechanics*, 6th edn. London, U.K.: E & FN Spon, 485pp.

11.2 Taylor, D.W. 1937. Stability of earth slopes. *Journal of the Boston Society of Civil Engineers* 24(3):337–386.

11.3 Gibson, R.E. and Morgenstern, N.R. 1962. A note on the stability of cuttings in normally consolidated clay. *Geotechnique* 12(3):212–216.
11.4 TRB. 1996. *Landslides—Investigation and Mitigation*, Turner, A.K. and Schuster, R.L., eds. US Transportation Research Board Special Report 247. Washington, DC: Transportation Research Board, US National Academy of Sciences, 673pp.
11.5 Lambe, T.W. and Whitman, R.V. 1979. *Soil Mechanics*, S.I. Version. Toronto, Ontario, Canada: John Wiley & Sons, 553pp.
11.6 Fay, L., Akin, M., and Shi, X. 2012. *Cost-Effective and Sustainable Road Slope Stabilization and Erosion Control, a Synthesis of Highway Practice.* NCHRP Synthesis 430. Washington, DC: Transportation Research Board, 70pp.
11.7 Rodriguez, A.H., del Castillo, H., and Sowers, G.F. 1988. *Soil Mechanics and Highway Engineering.* Clausthal-Zellerfield, Federal Republic of Germany: Trans Tech Publications, 843pp.

Chapter 12

Geosynthetics

12.1 BACKGROUND

Geosynthetics—filter cloth, filter fabrics, geotextiles, geogrids, geonets, geomembranes, geowebs, geocomposites—exploded into geotechnical engineering in the 1970s. The North American market alone went from under 8×10^6 m^2 (10×10^6 yd^2) sold in 1976 to over 500×10^6 m^2 (600×10^6 yd^2) in 1992 [12.1]. The global market was estimated at approximately 2×10^9 m^2 (2.4×10^9 yd^2) in 2010, and is forecast to be in excess of 5×10^9 m^2 (6×10^9 yd^2) by 2018.* The dramatic development of a market for these new products for earthworks occurred for the following reasons [12.1]:

- They are products manufactured under factory conditions, including quality control
- They can be installed rapidly
- They generally replace or reduce the use of natural materials (aggregate, for example)
- They are generally cost-competitive with the natural materials they replace

When first introduced, geosynthetics were a new, radical departure from typical geotechnical engineering. The incentive to use them was high, but the risk of failure was also high. It has taken a couple decades for research to provide the required reliable new design methods. Landmark gatherings for the dissemination of research took place with international conferences, the first being held in Paris in 1977, followed by conferences in Las Vegas (1982), Vienna (1986), the Hague (1990), Singapore (1994), Atlanta (1998), Nice (2002), Yokohama (2006), Guarujá (2010), and Berlin (2014). At least two learned journals are published, and countless regional symposia are now held.

The first geosynthetics were used for filtration, but soon other functions were developed.

* Geosynthetics Market, 2014, http://www.marketsandmarkets.com/PressReleases/geosynthetics.asp, visited January 2, 2014.

12.2 FUNCTIONS

Geosynthetics are placed in geotechnical structures to provide one or more of the following functions:

- Separation
- Reinforcement
- Filtration
- Drainage
- Liquid containment

Containment is not relevant to LVR; the first four are relevant, and are discussed in the following.

12.2.1 Separation

The old adage, "put half a metre of good gravel down on half a metre of mud, and you'll end up with a metre of mud" speaks to the separation function of geosynthetics. When coarse-grained soil is placed on fine-grained soil, the particles of the fine-grained soil tend to find their way into the voids of the coarse-grained soil, and the particles of the coarse-grained soil tend to punch into the fine-grained soil. The mixture that results is weaker than the uncontaminated coarse-grained soil. Placing the appropriate geosynthetic between the two materials prevents their mixing, resulting in a stronger structure.

12.2.2 Reinforcement

Geosynthetics, which are generally strong and stiff in tension, can provide the tensile resistance soil lacks. This reinforcement function can be fulfilled by the geosynthetic in several ways:

- The geosynthetic can deform under load under a road base, providing tensioned membrane reinforcement by acting like the fabric of a trampoline.
- Reinforcement can be provided by the geosynthetic resisting the shear stresses arising in the soil under a footing or embankment.
- Geosynthetics used as the tie back elements of a wall provide reinforcement in anchorage.

12.2.3 Filtration

Beyond just keeping two different soils from mixing, as in the separation function, it may be that the geosynthetic must also allow water to pass through, relieving a build-up of pore water pressure. This is the filtration function. The geosynthetic must be porous enough and have holes large enough to allow the easy flow of water, while still having holes small enough to retain the fine particles of soil. It must simultaneously have high permeability (or *permittivity* if the thickness of the geosynthetic is accounted for) and high soil retention.

12.2.4 Drainage

If thick enough and porous enough, a geosynthetic can provide a drainage capability within the plane of the geosynthetic, for example carrying water to the drainage ditches at the side of a road. Again permeability, this time in the plane of the geosynthetic (called *transmissivity*), and soil retention are important.

12.3 GEOSYNTHETIC TYPES AND MANUFACTURE

The geosynthetics used in LVR applications are usually one of three types: wovens, non-wovens, and grids. Beyond composites of these three types, knits and nets are available, and the development of new types continues.

Woven geotextiles are produced on conventional looms (Figure 12.1). In the simplest weave, every other yarn in the longitudinal direction, called the warp or machine direction, is alternately lifted and dropped by comb-like parts of the loom called the reeds. A shuttle carries the weft or cross-machine-direction yarns from one side to the other through the gap between the machine direction yarns, once for each movement of the reeds. The reeds flip the machine-direction yarns with each pass of the shuttle from side to side. Various weave patterns are achieved by changing the pattern of the lifting and dropping of the machine-direction yarns. The weaving patterns have a direct effect on the properties of the geotextile.

The yarns are produced in a wide variety of forms [12.1]:

- *Monofilaments* similar to fishing line can be extruded by spinnerets, with the molten polymer emerging from a die, like toothpaste from hundreds of toothpaste tubes.
- Solid sheets of polymer can be cast and slit into narrow tapes with knives or air jets, resulting in *slit-film monofilaments.*
- The monofilaments can be twisted together to make *multifilament yarns.*
- The slit-film monofilaments can be twisted together to make *slit-film multifilament yarns.*
- Monofilament yarns can be chopped into short fibers—*staples*—and spun together to make *staple yarns.*

Figure 12.1 Loom producing woven geogrid. Yarns in the background pass through the reeds in the middle distance, emerging as a woven product in the foreground. (Photo courtesy Tencate Geosynthetics Americas.)

Non-woven geotextiles are produced with a completely different process. There are many variations possible, but basically the production of non-woven geotextiles involves fiber production, web formation, web bonding, and final treatment [12.1]. Fibers are extruded by spinnerets, cooled, and stretched. The stretching aligns the long-chain molecules of the polymer, adding strength and decreasing creep. The fibers are laid down in random orientation on a moving belt. The fibers of the resultant web are then bonded with a thermal, chemical, or mechanical treatment. In the melt-bonding treatment, they are heated enough to fuse the fibers together where they touch. In the chemical treatment, the geotextile web is impregnated with acrylic resin, bonding the fibers together. In the needle-punching mechanical treatment, plates with barbed needles are forced through the web, and then withdrawn, tangling the fibers (Figures 12.2 and 12.3). Final treatment may include calendaring, where hot rollers press the mat thinner and further fuse the fibers together.

Some grids are made by continuously extruding a molten sheet of polymer, punching it with rectangular or triangular holes, then stretching the remaining strands. The stretching aligns the long-chain molecules of the polymer, giving the grid strength and tensile stiffness. Other grids are produced with the weaving process (Figure 12.1).

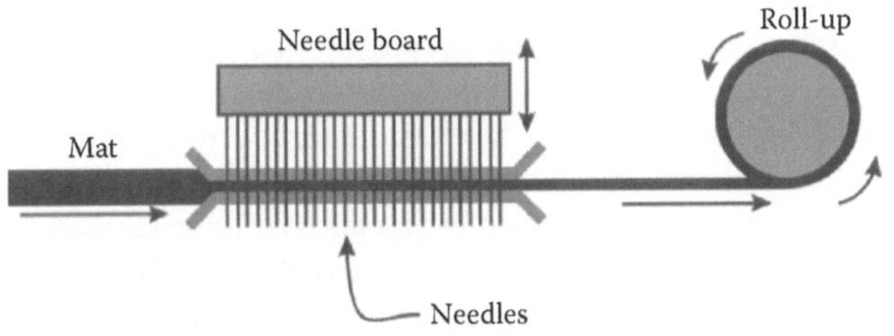

Figure 12.2 Needle punching nonwoven geotextile.

Figure 12.3 Needle boards (inverted). (Photo courtesy Tencate Geosynthetics Americas.)

12.4 GEOSYNTHETIC MATERIALS

The majority of geosynthetics are made from polymers (plastics), including the following [12.2]:

- Polypropylene
- Polyester
- Polyethylene
- Polyamide (nylon)

All are materials composed of long-chain molecules (polymer comes from *polumeros*, Greek for "having many parts" [12.3]). The molecules have chains of repeating units—mers—and the chain structure imparts tensile strength and stiffness. Breaking the chains causes a weakening of the material. Exposure to ultraviolet light can break the chains, hence the need for careful packaging and storage of rolls of a geosynthetic to shield them from sunlight, and subsequent burying of the geosynthetic in its road or reinforced steep slope, imperative to avoid a loss of strength.

General properties of the common polymers used in geosynthetics are shown in Table 12.1. Of particular practical significance is the specific gravity. Whether or not a geosynthetic floats has construction implications.

12.5 TESTS

Tests for quality control had to be devised quickly as geosynthetics developed. Naturally many were borrowed from the conventional textile industry. Others had to be developed from scratch and refined. Numerous stakeholders took on the role, and by 1994 Koerner [12.2] identified at least 32 agencies and institutes developing standards (Table 12.2). By 2012 when Koerner published Reference 12.1, the list was dropped. By that time it was estimated that over 300 test standards had been developed.

This book will concentrate on the testing standards published by ASTM and ISO. First, tests of geotextiles will be covered, and then tests of geogrids will follow.

12.5.1 Tests on geotextiles

The relevant tests on geotextiles can be grouped as follows:

- Tests for physical properties:
  - Specific gravity, mass per unit area, thickness, stiffness
- Tests for mechanical properties:
  - Compressibility, tensile strength (grab test, wide strip test, confined test), seam strength, fatigue strength, burst strength, tear, impact, puncture, friction, pullout (anchorage)
- Tests for hydraulic properties:
  - Porosity, percent open area, apparent opening size (or equivalent opening size), permittivity (cross-plane permeability), transmissivity (in-plane permeability), soil retention
- Tests for endurance properties:
  - Installation damage, creep (constant stress), stress relaxation (constant strain), stress (creep) rupture, abrasion, long-term flow (clogging), gradient ratio (clogging), hydraulic conductivity (clogging)

Table 12.1 Physical properties of polymers used in geosynthetics

Polymer	Breaking tenacity (g/denier[a])		Specific gravity	Standard moisture regain (%)	Coefficient of thermal expansion		Melting range	
	Standard	West			($\times 10^{-5}$/°C)	($\times 10^{-5}$/°F)	(°C)	(°F)
Polyethylene								
High density	—	—	0.96	2.0	22.5	12.5	—	—
Polypropylene								
Filament and staple	4.8–7.0	4.8–7.0	0.91	3.0	11	6.2	160–179	325–335
Polyester								
Regular-tenacity filament	4.0–5.0	4.0–5.0	1.22 or 1.38	0.4 or 0.8	7.6–8.6	4.2–4.8	250–290	480–550
High-tenacity filament	6.3–9.5	6.2–9.4	1.22 or 1.38	0.4 or 0.8	7.6–8.6	4.2–4.8	250–290	480–550
Regular-tenacity staple	2.5–5.0	2.5–5.0	1.22 or 1.38	0.4 or 0.8	7.6–8.6	4.2–4.8	250–290	480–550
High-tenacity staple	5.0–6.5	5.0–6.4	1.22 or 1.38	0.4 or 0.8	7.6–8.6	4.2–4.8	250–290	480–550
Nylon								
Nylon 66: regular-tenacity filament	3.0–6.0	2.6–5.4	1.14	4.0–4.5	9.9	5.5	≈260	≈500
Nylon 66: high-tenacity filament	6.0–9.5	5.0–8.0	1.14	4.0–4.5	9.9	5.5	≈260	≈500
Nylon 66: staple	3.5–7.2	3.2–6.5	1.14	4.0–4.5	9.9	5.5	≈260	≈500
Nylon 6: filament	6.0–9.5	5.0–8.0	1.14	4.5	9.0	5.0	210–220	414–428
Nylon 6: staple	2.5	2	1.14	4.5	9.0	5.0	210–220	414–428

Source: Koerner, R.M., *Designing with Geosynthetics*, 3rd edn., Prentice-Hall Inc, Englewood Cliffs, NJ, 783pp., 1994.

[a] Denier is a measure of yarn or filament mass, defined as the mass of 9,000 m (29,500 ft) of yarn or filament.

Table 12.2 Agencies and institutes developing geosynthetic test standards, as of 1994

	Agency of institution
International	International Organization for Standards (ISO)
	Permanent International Association of Road Congresses (PIARC)
	European Disposables and Non-wovens Associated (EDANA)
	Permanent International Association of Navigation Congresses
	Geosynthetic Research Institute (GRI) of Drexel University
Belgium	Ministry of Public Works
Canada	Canadian General Standards Board (CGSB)
Finland	Technical Research Centre of Finland
France	Comité Français des Géotextiles et Géomembranes
France	L'Association Français de Normalisation (AFNOR)
France	Réunion Internationale des Laboratoires d'Essais et de Recherche sur les Matériaux et les Constructions (RILEM)
Germany	Bundesanstalt für Wasserbau
Germany	Deutsche Gesellschaft für Erd- und Grundbau
Germany	Franzius Institut (für Wasserbau und Küstingenieurswesen der Universität Hannover)
Germany	German Standards Committee for Geotextiles (DIN)
Italy	Consiglio Nazionale Richerche (CNR)
Italy	ENEL/CRIS
Italy	Unitex
Korea	Korea Highway Corp.
Netherlands	Nederlands Normalisatie Instituut (NNI)
Norway	Norwegian Road Research Laboratory (NRRL)
Portugal	João de Matos Roasa
South Africa	South Africa Bureau of Standards
Sweden	Swedish Geotechnical Institute
Sweden	The Swedish National Road Administration
Switzerland	Schweizerisches Verband der Geotextilfachleute
United Kingdom	British Standards Institution (BSI)
United Kingdom	The Institution of Civil Engineers (ICE)
United Kingdom	Transport and Road Research Laboratory (TRRL)
United States	American Association of State Highway and Transportation Officials (AASHTO)
United States	The American Society for Testing and Materials
United States	U.S. Army Corps of Engineers, Waterways Experiment Station

Source: Koerner, R.M., *Designing with Geosynthetics*, 3rd edn., Prentice-Hall Inc., Englewood Cliffs, NJ, 783pp., 1994.

- Tests for degradation resistance:
  - Sunlight (ultraviolet) degradation, temperature degradation, oxidation degradation, hydrolytic degradation, chemical degradation, biological degradation, radioactive degradation, aging

A handful of the more commonly applied tests will be discussed here, to facilitate the writing of specifications. For further information, note that Koerner [12.1,12.2] provides a full treatment of each test listed above, or the individual ASTM or ISO standards can be consulted.

12.5.1.1 Mass per unit area (ASTM D5261, ISO 9864)

What is often referred to as the "weight" of the geotextile is actually the mass per unit of area. The value is reported in g/m^2, or oz/yd^2. The specimen should not be under any tension when the length and width are measured. Geotextiles generally range from 150 to 750 g/m^2 (4.5 to 22 oz/yd^2) although geotextiles in excess of 2000 g/m^2 (60 oz/yd^2) have been used [12.1].

12.5.1.2 Tensile strength (ASTM D751, D4595, D4632; ISO 10319)

"Strength" is the maximum tensile load at the point of failure, usually defined in units of kN/m (lb/ft). It should be reported in both the machine direction (warp direction) and cross-machine direction (weft direction) as the two will probably not have the same value. Often the test data is reported as the strength at a given strain, or presented on a graph of strength versus elongation or strain.

In the test, the geotextile specimen is clamped in parallel jaws, and pulled at a constant rate in a testing machine. There is a trade-off to be made between getting an accurate assessment of the strength of the geotextile, which will neck down during the test, and the size and cost of the test specimen. As a result, there are a number of different standard widths for the test, with corresponding length-to-width ratios. The common specimen sizes are given in Table 12.3.

The grab test is a special test, intended to mimic the stretching effect on the geotextile imposed by individual stones in the aggregate gripping the geotextile and spreading under load. The clamps for the grab test (25 mm or 1 $in.^2$) are not equal to the width of the specimen (100 mm or 4 in.). The grab test is a very common test reported on geotextile specification sheets, and is used in the assessment of geotextile survivability (Section 12.5.1.9).

Eventually one comes to the end of the roll of geotextile, and there may be a need for a seam to connect to the next roll—seam strength becomes an important consideration. The tensile strength of the seams can be less than the tensile strength of the geotextile itself, and the design of the installation must take this reduced strength into consideration. Seam strength is conveniently expressed as seam efficiency, Equation 12.1.

Table 12.3 Geotextile strength test specimen details

Specification	Specimen width (mm)	Specimen width (in.)	Specimen length between inner faces of clamps (mm)	Specimen length between inner faces of clamps (in.)	Remarks
ASTM D4632	100	4	75	3	Clamps are 25 mm square, not equal to specimen width
ASTM D751	25 or 50	1 or 2	n/a	n/a	Narrow-strip test
ASTM D4595, ISO 10319	200	8	100	4	Wide-width test
No specification	1000	—	500	—	Not a standard test arrangement

Source: Koerner, R.M., *Designing with Geosynthetics*, 6th edn., Vol. 1, Xlibris, Bloomington, IN, 2012.

$$E = \frac{T_{seam}}{T_{geotextile}} \times 100 \tag{12.1}$$

where

E is the seam efficiency (percent)
T_{seam} is the tensile strength of the seam (kN/m, or lb/ft)
$T_{geotextile}$ is the wide-width tensile strength of the geotextile (kN/m, or lb/ft)

Note that for higher geotextile tensile strengths, the seam efficiency tends to be lower [12.2], and can be as little as 50%, having a significant impact on design.

12.5.1.3 Tear strength (ASTM D4533, D1424; ISO 13434)

Tear strength is another test used in the assessment of geotextile survivability (Section 12.5.1.9).

In the trapezoidal tear test, a trapezoidal specimen is held between parallel clamps, leaving it taut at one side, and slack at the other. A 15 mm (0.59 in.) cut is made in the taut end, and the specimen loaded to failure, progressively failing across its width.

In the tongue tear test, a 75 mm (3 in.) by 200 mm (8 in.) specimen with a 75 mm (3 in.) initiation cut is placed in a tension testing machine, with the cut ends in the clamps. Tension is applied, and the specimen tears, starting at the root of the initiation cut [12.1]. The failure tension is reported as the tear strength. In this test arrangement, the yarns work together, resulting in higher tear strengths than the trapezoidal tear test gives.

12.5.1.4 Puncture resistance (ASTM D4833, ASTM D6241, ISO 12236)

Puncture resistance is also a test used to determine geotextile survivability (Section 12.5.1.9). The test is intended to mimic the penetration of rocks, branches, and roots into the geotextile under static load.

In one arrangement of the test, the specimen is firmly clamped over the end of a 45 mm (1.75 in.) I.D. hollow cylinder, and an 8 mm (0.315 in.) rod pushed into it using a compression testing machine. The shape of the end of the rod—hemispherical, flat, or beveled—has an effect on the test results. In the standard, the end of the rod is flat with a 0.8 mm (0.032 in.) 45° bevel.

The small diameter of the rod, relative to the yarn spacing of the geotextile, is a concern. An improvement is a puncture test where a standard California bearing ratio (CBR) loading cylinder and an empty CBR mould are used. The cylinder is 50 mm (1.95 in.) in diameter, and the mould is 150 mm (6 in.) in inside diameter. The cylinder has a 0.80 mm (0.032 in.) bevel cut at 45°. The fabric is stretched taut over the end of the CBR mould, and the cylinder pushed into it, as is done in the CBR test. The load at failure is reported.

12.5.1.5 Percent open area

The percent open area test (POA) is applicable only to woven monofilament geotextiles [12.1]. The ratio of open area of the geotextile to the overall area is determined in the test, giving a measure of the water flow capacity through the geotextile. To carry out the test, light is projected through the geotextile onto a large sheet of paper or cardboard. The magnified open spaces are traced, and their area determined with a planimeter or squared paper. Woven monofilament geotextiles vary in POA from essentially closed at 0% to extremely open at about 36% [12.1]. The general range is from 6% to 12%. The overlapping yarns of non-woven geotextiles preclude non-woven geotextiles from this test. The apparent opening size test fulfils the need.

12.5.1.6 Apparent opening size (ASTM D4751)

The apparent opening size (AOS or equivalent opening size, EOS, the two are equivalent) of the geotextile is determined by sieving glass beads, with a known grain size distribution curve, through the geotextile. The AOS is the size for which 5% pass through the geotextile. This corresponds to the 95% opening size, O_{95} of the geotextile. Test repeatability is difficult to achieve with non-woven geotextiles as the beads get trapped in the voids in the geotextile. There are numerous other problems with the test, but it persists [12.1].

The filtration opening size (FOS) test is similar to the AOS/EOS test, except that rather than dry sieving, wet sieving using hydrodynamic forces is employed [12.6]. Above 150 μm, AOS/EOS and FOS results are similar. FOS is recommended for geotextiles below this value [12.4].

12.5.1.7 Permittivity (ASTM D4491, ASTM D5493, ISO 11058)

In cases where the geotextile acts as a filter, its cross-plane permeability is of interest. Because the geotextile has thickness, and that thickness may change under load, the permeability per unit of thickness—its permittivity—becomes an important design parameter. Permittivity is derived from Darcy's equation:

$$q = k_n i A = k_n \frac{\Delta h}{t} A \tag{12.2}$$

$$k_n = \frac{qt}{\Delta h A}$$

$$\psi = \frac{k_n}{t} = \frac{q}{\Delta h A}$$

where

- q is the flow rate (m^3/s or ft^3/s)
- k_n is the permeability (m/s or ft/s)
- i is the hydraulic gradient (m/m or ft/ft)
- Δh is the total head loss (m or ft)
- t is the thickness of geotextile (m or ft)
- A is the area of geotextile (m^2 or ft^2)
- ψ is the permittivity (s^{-1})

The test is usually done in a constant head test permeameter, with the geotextile specimen held where the soil would normally be placed, run at a number of head values, and the permittivity calculated as in Equation 12.2. It can also be run as a falling head test, in which ψ is calculated in Equation 12.3.

$$\psi = 2.3 \frac{a}{A \Delta t} \log_{10} \frac{h_o}{h_f} \tag{12.3}$$

where

- a is the area of the supply pipe (m^2 or ft^2)
- A is the area of the geotextile (m^2 or ft^2)

Δt is the time taken for the head to fall from h_o to h_f (s)
h_o is the total head at the beginning of the test (m or ft)
h_f is the total head at the end of the test (m or ft)
ψ is the permittivity (s^{-1})

Typical permittivity values range over several orders of magnitude, from 0.02 to 2.2 s^{-1}. Corresponding permeabilities range from about 8×10^{-6} to 2×10^{-3} m/s (30×10^{-6} to 7×10^{-3} ft/s) [12.1].

Some geotextiles, particularly the thick non-wovens, change thickness with pressure. Therefore, a permittivity test for the geotextile under load has been developed. The same apparatus is used, with ceramic balls applying the pressure.

12.5.1.8 Soil retention (ASTM D5142)

The issue of geotextile soil retention arises in two ways: for underwater turbidity curtains, and for above-ground silt fences. The latter will be discussed here.

In the test, a slurry of silty sand is poured into a flume box 1.2 m (47 in.) long, tipped at an 8% slope, and the water and fines flow through the geotextile fixed across the lower end. The slurry flow rate through the geotextile and the retention efficiency of the geotextile are reported [12.1].

12.5.1.9 Survivability

In many cases, the handling and installation of the geosynthetic is potentially more damaging than conditions during the service life in the road, embankment, or slope. As well as service, the geosynthetic needs to survive construction. Factors such as the condition of the subgrade the geosynthetic will be placed on, the ground pressure of the equipment to run on it, the thickness of the first lift of the backfill placed on it, and the particle size and shape (angularity) of the backfill all have an impact on the survivability requirements.

To address survivability, a two-step approach is taken. First the construction circumstances are rated, and then based on that rating, a set of minima for the grab strength, puncture resistance and tear strength are identified. Tables 12.4 and 12.5 present the ratings and the minimum test values, respectively. If the requirements given in Table 12.5 exceed those of a specific design criterion (Section 12.6), the geotextile must meet the requirement given in Table 12.5.

12.5.2 Tests on geogrids

Geogrids being used for just one function, reinforcement, only a handful of tests are performed on them. Three are discussed below.

Table 12.4 Level of survivability

Site soil CBR during installation[a]	*<1*		*1–2*		*>3*	
Equipment ground pressure (kPa)[g]	*>35*	*<35*	*>35*	*<35*	*>55*	*<55*
Compacted cover thickness (mm)[b]						
100[c,d]	NR[f]	NR	High	High	Medium	Medium
150[e]	NR	NR	High	High	Medium	Low
300	NR	High	Medium	Medium	Medium	Low
450	High	Medium	Medium	Medium	Medium	Low

Source: CGS, *Canadian Foundation Engineering Manual*, 4th edn., Canadian Geotechnical Society (CGS) c/o BiTech Publisher Ltd., Richmond, British Columbia, Canada, 488pp., 2006. © Canadian Geotechnical Society (CGS), used with permission.

a Assume saturated CBR unless it can be controlled.
b Maximum aggregate size not to exceed one-half the compacted thickness.
c For unpaved roads with ADT <200 veh/day.
d The 100 mm cover is limited to existing road bases and is not intended for use in new construction.
e Maximum aggregate size <30 mm.
f NR = not recommended, high = high survivability requirements, etc.
g These values have been amended from those shown in [12.4] (350 and 550): manufacturers provide specifications for LGP equipment as about 35 kPa.

12.5.2.1 Wide-width tensile strength (ASTM D6637, ISO 10319)

As was the case for geotextiles (Section 12.5.1.2), a wide specimen is clamped in a testing machine, and pulled to failure. The width and length of the specimen, strain rate, and method of measuring deformation must all be agreed by the parties involved [12.1]. Test results reported include tensile strength, strain at failure, and the tensile strength at various strains prior to failure, for example, 1%, 2%, and 5%.

The tensile strength of geogrids generally falls between that of conventional geotextiles and geotextiles specifically designed for high strength.

12.5.2.2 Tension creep (ASTM D5262, ISO 13431)

An important issue with geogrids is their elongation under constant load—creep. The molecular structure of the polymer in the geogrid, plus the stress level, time, temperature, and other environmental factors all influence the amount of creep. A maximum amount of creep strain, say 10%, can be adopted, and the stress level found that matches that amount of creep strain. This sets the limiting stress value that can be permitted in the actual field application.

Table 12.5 Physical property requirements for survivability

	Grab strength ASTM D4632 (N)		Puncture resistance ASTM D4833 (N)		Tear strength (ASTM D4533 (N)	
Geotextile survivability level	<50% Geotextile elongation	>50% Geotextile elongation	<50% Geotextile elongation	>50% Geotextile elongation	<50% Geotextile elongation	>50% Geotextile elongation
High (Class 1)	1400	900	500	350	500	350
Moderate (Class 2)	1100	700	400	250	400	250
Low (Class 3)	800	500	300	180	300	180
Additional requirements			**Test method**			
Apparent opening size			ASTM D4751			
<50% soil passing 0.075 mm sieve: AOS <0.6 mm						
>50% soil passing 0.075 mm sieve: AOS <0.3 mm						
Permeability			ASTM D4491			
k of the geotextile >*k* of the soil (permeability = permittivity x the nominal geotextile thickness)						
Ultraviolet light degradation			ASTM D4355			
at 500 h of exposure, at least 50% strength retained						
Geotextile acceptance			ASTM D4759			

Source: CGS, *Canadian Foundation Engineering Manual*, 4th edn., Canadian Geotechnical Society (CGS) c/o BiTech Publisher Ltd., Richmond, British Columbia, Canada, 488pp., 2006. © Canadian Geotechnical Society (CGS), used with permission.

12.5.2.3 Anchorage (pullout test)

To provide the required reinforcement of an installation, the geogrid needs to develop the required tension. It does so via anchorage somewhere in the soil structure, thus anchorage strength becomes an issue. The anchorage is generated by the surrounding soil particles striking through the apertures in the grid, and indeed, this is precisely why geogrids are formed the way they are, with large apertures. Sarsby [12.5] as reported in Koerner [12.1] recommends that a geogrid with the aperture dimension given in Equation 12.4 be selected.

$$B_{GG} \geq 3.5 d_{50} \tag{12.4}$$

where
B_{GG} is the minimum geogrid aperture width (mm or in.)
d_{50} is the average particle size of the backfill soil (mm or in.)

The anchorage force is developed from shear resistance on the top and bottom of the longitudinal and transverse strands, and from the passive resistance against the transverse ribs [12.1], enhanced by the strikethrough. Koerner [12.1] describes an anchorage test apparatus consisting of a large box approximately 1.9 m (75 in.) long, in which the geogrid specimen is placed horizontally. The geogrid protrudes through a slit in the side of the box, and is connected to a pulling mechanism. The box is backfilled with the prototypical soil that is loaded vertically. As the geogrid is pulled, the force required is measured, together with its extension. The pullout resistance is reported as the peak force divided by the area of the geogrid in the box.

12.5.3 Minimum average roll value: MARV

The value of any property (e.g., strength) will vary with a geosynthetic lot, and between lots. Any sampling of the geosynthetic and reporting of test results must take into account the statistical variation. To take a mean value for a lot can be very unrealistic because a lot may be huge compared to the amount of geosynthetic used for the project. The mean value may be so high as to be unsafe for the project. On the other hand, taking the minimum value for a specification requirement will also be unrealistically pessimistic of geosynthetic performance.

As a result, the minimum average roll value, MARV, is often referred to in specifications. This is the minimum value of all the rolls used in a project or specific part of a project. It is less than the mean lot value, and more than the minimum value and represents a reasonable minimum value for the specification.

12.6 DESIGNS

12.6.1 Applications relevant to LVR

While there are scores of applications for geosynthetics, a handful particularly relevant to LVR is covered here. For other applications such as retaining walls, references such as those by Koerner [12.1,12.2] and the Canadian Geotechnical Society [12.4] can be consulted.

12.6.2 Designing with geotextiles

12.6.2.1 Design methodology

Designs can be developed either by specification or by calculation from engineering principals [12.1]. If designed by specification, the minimum values for the geotextile's properties are selected from tables developed through experience. Otherwise, the values of properties relevant to the geotextile's function are calculated directly for the particular conditions of the design. In the following sections, the designs by specification produced by the American Association of State Highway and Transportation Officials (AASHTO) first are provided, followed by designs by calculation.

12.6.2.2 Designs by specification

AASHTO's [12.6] approach is based on the division of geotextiles into three classes:

- *Class 1*: for installations where survivability requirements are severe
- *Class 2*: for installations with typical survivability requirements. This class is assigned if no site-specific information is available
- *Class 3*: for installations with minimal survivability requirements

The required properties for the selection of a geotextile are based on the application the geotextile is to fulfill. The applications identified are separation, filtration, stabilization, permanent erosion control, and temporary silt fencing. Tables 12.6 through 12.10 provide the requirements for permittivity, apparent opening size, and ultraviolet light stability. The tables also indicate which survivability class the geotextile falls under. The required tensile strength for each class is given in Table 12.11.

Geotextiles for a given application must meet the requirements of the relevant table of Tables 12.6 through 12.10, and the required strength

Table 12.6 AASHTO M288 specification requirements for separation geotextiles

Property	*Test method*	*Units*	*Requirements*
Geotextile class (see Table 12.11)			Class 2
Permittivity	ASTM D4491	s^{-1}	0.02
Apparent opening size (max average roll value)	ASTM D4751	mm	0.60
Ultraviolet light stability (retained strength)	ASTM D4355	%	50 after 500 h exposure

Source: AASHTO, *Standard Specification for Geotextile Specification for Highway Applications, M288*, American Association for State Highway and Transportation Officials (AASHTO), Washington, DC, 2006.

Table 12.7 AASHTO M288 specification requirements for filtration geotextiles

			Requirements		
			Percent of in situ soil finer than 0.075 mm		
Property	*Test method*	*Units*	*<15*	*15–50*	*>50*
Geotextile class (see Table 12.11)			Class 2		
Permittivity	ASTM D4491	s^{-1}	0.5	0.2	0.1
Apparent opening size (max average roll value)	ASTM D4751	mm	0.43	0.25	0.22
Ultraviolet light stability (retained strength)	ASTM D4355	%	50 after 500 h exposure		

Source: AASHTO, *Standard Specification for Geotextile Specification for Highway Applications, M288*, American Association for State Highway and Transportation Officials (AASHTO), Washington, DC, 2006.

Table 12.8 AASHTO M288 specification requirements for stabilization geotextiles

Property	*Test method*	*Units*	*Requirements*
Geotextile class (see Table 12.11)			Class 1
Permittivity	ASTM D4491	s^{-1}	0.05
Apparent opening size (max average roll value)	ASTM D4751	mm	0.43
Ultraviolet light stability (retained strength)	ASTM D4355	%	50 after 500 h exposure

Source: AASHTO, *Standard Specification for Geotextile Specification for Highway Applications, M288*, American Association for State Highway and Transportation Officials (AASHTO), Washington, DC, 2006.

Table 12.9 AASHTO M288 specification requirements for permanent erosion-control geotextiles

			Requirements		
			Percent of in situ soil finer than 0.075 mm		
Property	*Test method*	*Units*	*<15*	*15–50*	*>50*
Geotextile class (see Table 12.11)					
Woven monofilament geotextiles			Class 2		
All other geotextiles			Class 1		
Permittivity	ASTM D4491	s^{-1}	0.7	0.2	0.1
Apparent opening size (max average roll value)	ASTM D4751	mm	0.43	0.25	0.22
Ultraviolet light stability (retained strength)	ASTM D4355	%	50 after 500 h exposure		

Source: AASHTO, *Standard Specification for Geotextile Specification for Highway Applications, M288*, American Association for State Highway and Transportation Officials (AASHTO), Washington, DC, 2006.

shown in Table 12.11. For example, a woven geotextile to be used in a base stabilization role would be specified as a Class 1 geotextile (Table 12.8). Its permittivity, apparent opening size, and retained strength under ultraviolet light would need to be at least 0.05 s^{-1}, 0.43 mm (maximum average roll value), and at least 50% after 500 h exposure (Table 12.8). Its grab strength, sewn seam strength, tear strength and puncture resistance would need to be at least 1400, 1260, 500, and 500 N, respectively (Class 1, Table 12.11).

12.6.2.3 Allowable versus ultimate geotextile properties

The laboratory test results discussed in Section 12.5 are the ideal results, and thus give upper bounds to the properties. In actual service, the ideal conditions will not be met, and results will be lower. A margin of safety can be introduced in one of two ways [12.1]:

1. Apply an extremely high factor of safety at the end of the design.
2. Apply reduction factors on the individual laboratory test values to create design-specific values.

The latter is recommended.

Table 12.10 AASHTO M288 specification requirements for temporary silt fence geotextiles

			Requirements		
				Unsupported silt fence	
Property	*Test method*	*Units*	*Supported silt fence*	*Elongation >50% (Nonwoven)*	*Elongation <50% (Woven)*
Maximum postspacing		m	1.2	1.2	2.0
Grab strength					
Machine direction (warp)	ASTM D4632	N	400	550	550
Cross-machine direction (weft)	ASTM D4632	N	400	450	450
Permittivity	ASTM D4491	s^{-1}	0.05	0.05	0.05
Apparent opening size (max average roll value)	ASTM D4751	mm	0.60	0.60	0.60
Ultraviolet light stability (retained strength)	ASTM D4355	%	70 after 500 h exposure		

Source: AASHTO, *Standard Specification for Geotextile Specification for Highway Applications, M288*, American Association for State Highway and Transportation Officials (AASHTO), Washington, DC, 2006.

Table 12.11 AASHTO M288 geotextile strength specification requirements by geotextile class

			Class 1		*Class 2*		*Class 3*	
Property	*Test method*	*Units*	*Elongation <50%, (Woven)*	*Elongation >50% (Nonwoven)*	*Elongation <50%, (Woven)*	*Elongation >50% (Nonwoven)*	*Elongation <50%, (Woven)*	*Elongation >50% (Nonwoven)*
Grab strength	ASTM D4632	N	1400	900	1100	700	800	500
Sewn seam strength	ASTM D4632	N	1260	810	990	630	720	450
Tear strength	ASTM D4533	N	500	350	400	250	300	180
Puncture resistance	ASTM D4833	N	500	350	400	250	300	180
Permittivity	ASTM D4491	s^{-1}	Minimum requirements for permittivity, apparent opening size, and ultraviolet light stability are given in individual tables by geosynthetic function, Tables 13.9 through 13.13					
Apparent opening size	ASTM D4751	mm						
Ultraviolet stability	ASTM D4355	%						

Source: AASHTO, *Standard Specification for Geotextile Specification for Highway Applications, M288*, American Association for State Highway and Transportation Officials (AASHTO), Washington, DC, 2006.

For designs involving geotextile strength, the formulation in Equation 12.5 is used.

$$T_{allow} = T_{ult}\left(\frac{1}{RF_{ID} \times RF_{CR} \times RF_{CBD}}\right) = T_{ult}\left(\frac{1}{\Pi RF}\right) \quad (12.5)$$

where

T_{allow} is the allowable tensile strength (kN/m or lb/ft)
T_{ult} is the ultimate tensile strength (kN/m or lb/ft)
RF_{ID} is the reduction factor for installation damage (≥1) (unitless)
RF_{CR} is the reduction factor for creep (≥1) (unitless)
RF_{CBD} is the reduction factor for chemical and biological degradation (≥1) (unitless)
$\Pi\ RF$ is the accumulated value of the strength reduction factors (unitless)

Typical values for selected reduction factors [12.1] are given in Table 12.12.

For designs for water flow, the same approach can be taken, with Equation 12.6 providing the formulation of the allowable flow rate.

$$q_{allow} = q_{ult}\left(\frac{1}{RF_{SCB} \times RF_{CR} \times RF_{IN} \times RF_{CC} \times RF_{BC}}\right) = q_{ult}\left(\frac{1}{\Pi RF}\right) \quad (12.6)$$

where

q_{allow} is the allowable flow rate (m^3/s or ft^3/s)
q_{ult} is the ultimate flow rate (m^3/s or ft^3/s)
RF_{CC} is the reduction factor for biological clogging (≥1) (unitless)
RF_{BC} is the reduction factor for creep reduction of void space (≥1) (unitless)

Table 12.12 Recommended strength-reduction factors to determine allowable geotextile tensile strength

Application	*Installation damage, RF_{ID}*	*Creep, RF_{CR}*	*Chemical/biological degradation, RF_{CBD}*
Separation	1.1–2.5	1.5–2.5	1.0–1.5
Unpaved roads	1.1–2.0	1.5–2.5	1.0–1.5
Embankments	1.1–2.0	2.0–3.5	1.0–1.5
Silt fences	1.1–1.5	1.5–2.5	1.0–1.5

Source: Koerner, R.M., *Designing with Geosynthetics*, 6th edn., Vol. 1, Xlibris, Bloomington, IN, 2012. © Robert M. Koerner, used with permission.

Table 12.13 Recommended flow-reduction factors to determine allowable geotextile flow rate

Application	*Soil clogging and blinding, RF_{SCB}*	*Creep reduction of voids, RF_{CR}*	*Intrusion into voids, RF_{IN}*	*Chemical clogging, RF_{BC}*	*Biological clogging, RF_{BC}*
Underdrain filters	5.0–10	1.0–1.5	1.0–1.2	1.2–1.5	2.0–4.0
Erosion control filters	2.0–10	1.0–1.5	1.0–1.2	1.0–1.2	2.0–4.0

Source: Koerner, R.M., *Designing with Geosynthetics*, 6th edn., Vol. 1, Xlibris, Bloomington, IN, 2012. © Robert M. Koerner, used with permission.

RF_{IN} is the reduction factor for adjacent materials intruding into the geotextile's void space (≥1) (unitless)
RF_{CC} is the reduction factor for chemical clogging (unitless)
RF_{BC} is the reduction factor for biological clogging (unitless)
$\Pi\ RF$ is the accumulated value of the flow reduction factors (unitless)

Typical values for selected reduction factors [12.1] are given in Table 12.13.

12.6.2.4 Designing for separation with geotextiles

Two things stand out about the separation function of geotextiles: it is the most under-rated function, and many if not most applications involve separation [12.2]. The properties that are relevant to this important function are burst resistance, tensile strength, and puncture resistance.

A geotextile placed on soil, with a stone course placed on top of it, will tend to be pushed into the spaces between individual stones. The arrangement is modelled by the Mullen burst test for geotextiles, where the geotextile is pressured into a dome shape until it fails in radial tension. Equation 12.7 is the governing equation (see Koerner [12.1] or [12.2] for the full development).

$$FS = \frac{T_{allow}}{T_{reqd}} = \frac{p_{test}\, d_{test}}{(\Pi RF)\, p' d_v} \tag{12.7}$$

where

FS is the global factor of safety
T_{allow} is the allowable tension (kN or lb)
T_{reqd} is the tension actually required by the arrangement (kN or lb)
p_{test} is the burst test pressure, assumed equal to the pressure in the soil (kN/m^2 or $lb/in.^2$)
d_{test} is the diameter of the burst test apparatus, 1.2 in. (0.030 m)
$\Pi\ RF$ is the cumulative effect of the reduction factors (see Section 12.6.2.3) (unitless)

p' the stress at the geotextile's surface (kN/m^2 or lb/in.^{2})

d_v is the diameter of the voids in the stone course, taken to be 1/3 the average stone size (m or in.)

As an example, a geotextile with an ultimate burst pressure of 285 lb/in.^{2} (1970 kPa) is being checked as suitable for separation, in a road structure built of 2 in. (0.050 m) maximum size stone, subjected to a tire inflation pressure of 100 lb/in.^{2} (690 kN/m^2). The product of the partial reduction factors is calculated to be, say, 1.5. The global factor of safety is calculated to be (285 × 1.2)/(1.5 × 100 × 0.33 × 2) = 3.5, which is acceptable.

Consider two stones sitting on the geotextile. A third wedges between them, causing the first two stones to grip the geotextile and attempt to spread it. The situation is modeled by the grab test (Section 12.5.1.2). Equation 12.8 is the governing equation.

$$FS = \frac{T_{allow}}{T_{reqd}} = \frac{T_{test}}{(\Pi RF)p'd_v^2\ [f(\varepsilon)]} \tag{12.8}$$

where

FS is the global factor of safety (unitless)

T_{allow} is the allowable tension (kN or lb)

T_{reqd} is the tension actually required by the arrangement (kN or lb)

T_{test} is the grab test tension (kN or lb)

$\Pi\ RF$ is the cumulative effect of the reduction factors (see Section 12.6.2.3) (unitless)

p' is the pressure on the geotextile's surface (kN/m^2 or lb/in.^{2})

d_v is the diameter of the voids in the stone course, taken to be 1/3 the average stone size (m or in.)

$f(\varepsilon)$ is a function of the strain in the geotextile, shown to be 0.52 for the strain, estimated [12.2] to be 0.33 (unitless)

If in the road example the geotextile has a grab strength of 125 lb (556 N = 0.556 kN), what is the global factor of safety? The maximum stone diameter is 2 in. (0.05 mm), the cumulative reduction factors are 1.5 and the stress on the geotextile is 100 lb/in.^{2} (690 kPa). The global factor of safety is (125)/(1.5 × 100 × (0.33 × 2)^2 × 0.52) = 3.7, which is acceptable.

The geotextile in a road structure will be laid on the subgrade, which may have sticks, stumps, sharp rocks, or other things that may puncture it. Puncture test results (Section 12.5.1.4) are used to check the suitability of the geotextile. Equation 12.9 is the governing equation.

$$FS = \frac{F_{allow}}{F_{reqd}} = \frac{F_{test}}{\Pi RF} \times \frac{1}{p'd_a^2\, S_1\, S_2\, S_3} \tag{12.9}$$

where

FS is the global factor of safety (unitless)
F_{allow} is the allowable force to be resisted (lb)
F_{reqd} is the actual puncturing force actually acting in the arrangement (lb)
F_{test} is the puncture test result (lb)
$\prod RF$ is the product of the reduction factors (see Section 12.6.2.3) (unitless)
p' is the stress on the geotextile (lb/in.^{2})
d_a is the maximum stone size (in.)
S_1 is a protrusion factor, the height of protrusion into the geotextile by the object, divided by the average stone diameter (unitless)
S_2 is a scale factor to adjust the test diameter to the diameter of the actual puncturing object = $0.31/d_a$
S_3 is a shape factor, to adjust the flat probe of the puncture test to the actual shape of the puncturing object = $1 - A$, where A ranges from 0.8 for Ottawa sand, to 0.7 for pit run gravel, to 0.4 for crushed rock, to 0.3 for blasted rock

Note that the equation is not to be converted to metric units.

Continuing with the example, if the geotextile has a puncture resistance of 45 lb and the stress on the geotextile is 100 lb/in.^{2}, the accumulated reduction factor of 1.5, and factors S_1, S_2, and S_3 equal to 0.33, 0.155, and 0.6 respectively, the calculated global factor of safety is $(45/1.5) \times (1/(100 \times 2^2 \times 0.33 \times 0.155 \times 0.6)) = 2.4$, which is acceptable.

12.6.2.5 Designing embankment foundations with geotextiles

Embankments on soft soils can fail in any of at least four ways (Figure 12.4):

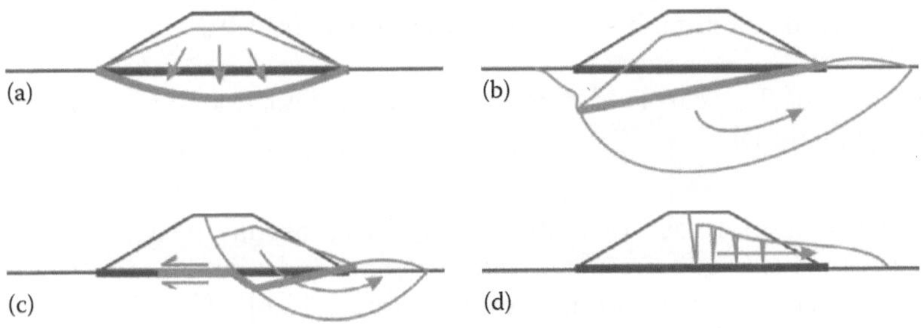

Figure 12.4 Failure modes of an embankment on a geosynthetic layer: (a) excessive deformation of foundation, (b) bearing failure of foundation soil, (c) global failure through embankment soil and foundation soil—either geosynthetic fails in tension or is pulled out of anchorage, and (d) lateral spreading of embankment soil on geosynthetic, or embankment soil and geosynthetic on foundation soil.

1. Excessive deformation of the foundation soil under the embankment
2. Bearing failure of the foundation soil under the embankment
3. Global failure through the embankment and foundation soil under the embankment
4. Lateral spreading of the embankment soil across the foundation soil

The provision of a geotextile at the embankment/foundation interface can reduce the risk of the third and fourth occurring, but not the first or second. In the case of excessive deformation of the foundation soil, the presence of a geotextile will not reduce the settlements nor differential settlements appreciably. However, the geotextile can act as a separator and filter, keeping the embankment soil uncontaminated by the foundation soil "on the way down."

The amount of elastic strain in the geotextile will limit the deformation in the embankment [12.1]. If there is too much strain, the embankment will be unserviceable. It has been recommended that the strain in the geotextile be limited to 10% [12.1], thus the geotextile modulus should be at least 10 T_{reqd}, Equation 12.10:

$$E_{reqd} = \frac{T_{reqd}}{\varepsilon_{max}} \tag{12.10}$$

$$E_{reqd} = \frac{T_{reqd}}{0.10}$$

$$E_{reqd} = 10T_{reqd}$$

where

E_{reqd} is the required geotextile modulus (kN/m or lb/ft)
T_{reqd} is the actual tension in the geotextile (kN/m)
ε_{max} is the maximum strain permitted, typically 10%

The value of T_{reqd} can be determined from numerical analysis (e.g., finite element analysis).

For the special case of embankments over fibrous peats, Rowe and Soderman [12.7,12.8] presented a design chart and a design table (see also the Canadian Foundation Engineering Manual [12.4]). These are provided in Figure 12.5 and Table 12.14, pertaining to embankments on fibrous peat over firm subgrades and over soft subgrades, respectively. It is assumed that the excess pore water pressure that develops in the peat layer does not exceed 0.34 $\Delta\sigma_v$, where $\Delta\sigma_v$ is the increment in vertical pressure due to the construction of the embankment. The reader is referred to the original papers for other assumptions.

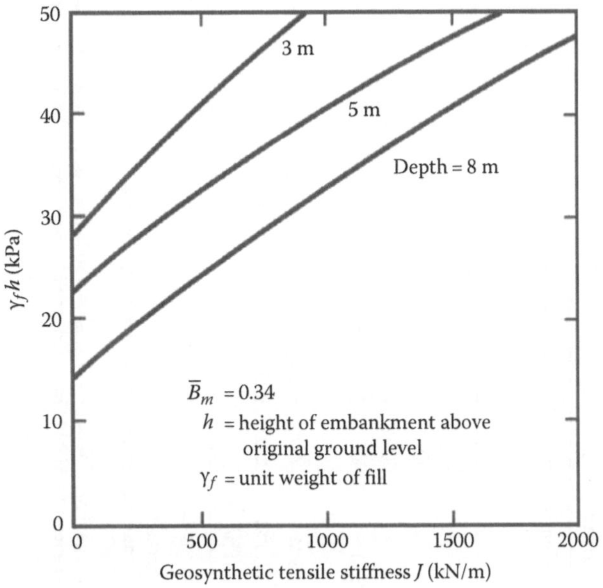

Figure 12.5 Design chart for embankment on peat underlain by a firm base. (After CGS, *Canadian Foundation Engineering Manual*, 4th edn., Canadian Geotechnical Society (CGS) c/o BiTech Publisher Ltd., Richmond, British Columbia, Canada, 488pp., 2006; Rowe, R.K. and Soderman, K.L., *Geotext. Geomembranes*, 2, 277. Copyright 1985 Canadian Geotechnical Society. Used with permission.)

In the case of bearing failure, the limiting height of the embankment is not affected by the geotextile. Thus conventional bearing capacity theory can be applied to the foundation soil.

Global stability is, however, influenced by the geotextile. Design is accomplished by trial and error, with a suite of failure arcs characterized by the location of their centers and their radii assumed and then analyzed. Failure on any assumed trial arc (Figure 12.6) is driven by the weight of the soil within the arc, and resisted by shear on the arc and by the tension in the geogrid. Equation 12.11 results from the moment equilibrium about the slip circle's center.

$$FS = \frac{c L_{arc} R + T_{allow}\, y}{W x} \qquad (12.11)$$

where

FS is the global factor of safety (should be ≥1.3) (unitless)
c is the cohesion (kN/m^2 or lb/ft^2)
L_{arc} is the length of the assumed failure arc (m or ft)
R is the radius of the assumed failure arc (m or ft)

Table 12.14 Design table for geosynthetic-reinforced embankments on peat over soft strata

Peat thickness (m)	Underlying stratum	Strength of underlying stratum, c_u (kN/m^2)	Maximum height of embankment above original ground level (m)			
			1.0	1.5	2.0	2.5
3	2 m clay	15	NRR[a]	NRR	500 (5%)[b]	1000 (6%)
		10	NRR	NRR	500 (6%)	1000 (6.5%)
3	2 m clay	7.5	NRR	150 (10%)	500 (8%)	1000 (7%)
		5	NRR	150 (14%)–500 (8%)[c]	2000 (4%)	PF[d]
3	2 m clay	2.5–5	150 (10%)	2000 (5.5%)	PF	PF
5	3 m clay	15	NRR	150 (14%)–500 (6.5%)	500 (9.5%)–1000 (5.5%)	1000 (6.5%)–2000 (4%)
		10	150 (5%)	500 (8.5%)–1000 (7.5%)	2000 (5.5%)	PF
5	3 m clay	7.5	500 (5%)	1000 (8.5%)–2000 (5%)	PF	PF

Source: Rowe, R.K. and Soderman, K.L., *Int. J. Geotext. Geomembranes*, 4(1), 65, 1986. © Geotextiles and Geomembranes, used with permission.

a NRR = no geosynthetic reinforcement required (separation and filtration requirements to be checked).

b 500 (5%) = geosynthetic with modulus J = 500 kN/m is recommended. Under assumed conditions, a maximum geosynthetic strain of approximately 5% is anticipated.

c 150 (14%) to 500 (8%): for the assumed conditions the embankment could be constructed with a geosynthetic with modulus J = 150 kN/m, but the expected strain of 14% is large. If conditions are likely to be as assumed, a higher modulus geosynthetic is recommended. J = 500 kN/m would result in approximately 8% strain under the assumed conditions.

d PF = potential failure for the assumed conditions: do not construct.

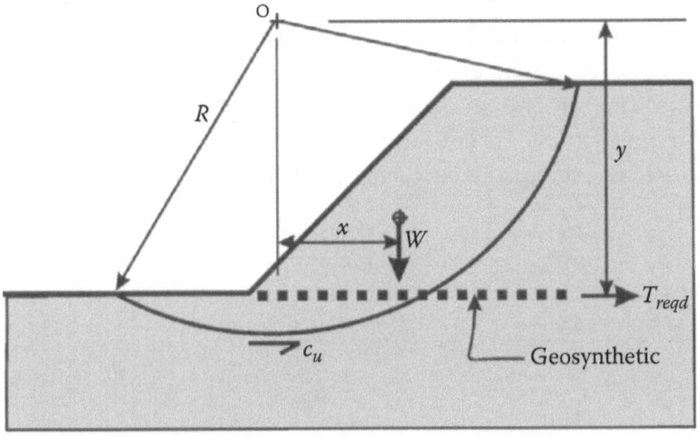

Figure 12.6 Trial failure arc for design of geosynthetic-reinforced embankment with steep slope: R = radius of trial slip circle, W = weight of slice, c_u = cohesion.

T_{allow} is the allowable tensile strength of the geotextile (kN/m or lb/ft)
y is the moment arm for the geotextile (m or ft)
W is the weight of soil within the failure arc (kN or lb)
x is the moment arm for W (m or ft)

If the embankment is thin, tension cracks may open up so it is sometimes assumed that no shear resistance is offered along the failure arc within the embankment soil.

It is rare that a slope is designed manually. Usually commercial software, which is based on the principles above, is used.

Associated with global instability is the possibility that the geotextile can be pulled out of the soil. With an adequate embedment length of the geotextile, this will not happen. Equation 12.12 provides the required embedment length.

$$L_{reqd} = \frac{T_{act}}{2E\,(c + \sigma_v \tan\phi)} \tag{12.12}$$

where

L_{reqd} is the required embedment length (m or ft)
T_{act} is the actual tension in the geotextile (kN/m or lb/ft)
E is the anchorage efficiency of the geotextile in the soil (unitless)
c is the cohesion of the soil (kN/m^2 or lb/ft^2)
σ_v is the average vertical stress on the geotextile, due to the soil above it (kN/m^2 or lb/ft^2)
ϕ is the angle of friction of the soil

It may be that the embankment soil slides laterally on the geotextile. Equations 12.13 and 12.14 are the governing equations [12.4], based on active pressure in the soil pushing outward, being resisted by shear on the top face of the geotextile, or on the bottom of the geotextile (which has failed in tension), respectively. Cohesionless soil and cohesive soil are assumed for the embankment and foundation soils respectively.

$$FS = \frac{L \tan \delta}{K_a H} \tag{12.13}$$

where

FS is the factor of safety on the embankment sliding on the geogrid (unitless)
L is the length of the spreading zone (m or ft)
δ is the angle of friction developed at the embankment/geotextile interface
K_a is the coefficient of active pressure = $\tan^2(45 - \phi/2)$ (unitless)
H is the height of the embankment (m or ft)

$$FS = \frac{2(Lc + T_{allow})}{K_a \gamma H^2} \tag{12.14}$$

where

FS is the factor of safety on the embankment and geotextile sliding on the foundation soil (unitless)
L is the length of the spreading zone (m or ft)
c is the adhesion on the geotextile/foundation soil interface (kN/m^2 or lb/ft^2)
T_{allow} is the allowable tensile strength of the geotextile (kN/m or lb/ft)
K_a is the coefficient of active pressure = $\tan^2(45 - \phi/2)$ (unitless)
γ is the unit weight of the embankment soil (kN/m^3 or lb/ft^3)
H is the height of the embankment (m or ft)

The design is considered adequate provided there are adequate factors of safety against each of the four failure modes—excessive deformation of the foundation soil under the embankment, bearing failure of the foundation soil, global failure through the embankment and foundation soil, and lateral spreading of the embankment soil across the foundation soil—and any other failure modes peculiar to the specific project. The need for separation and filtration between the embankment and the foundation soil should also be checked.

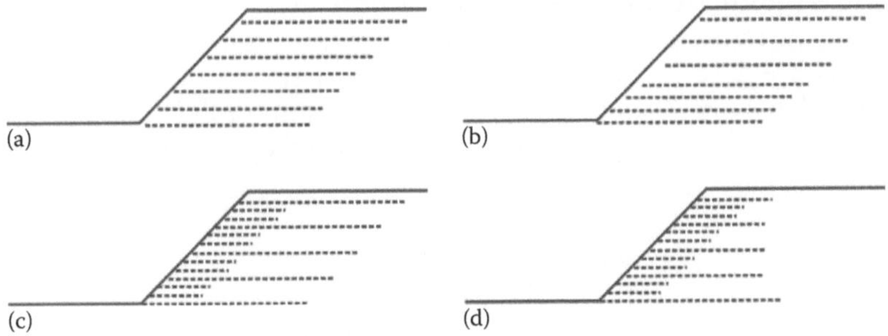

Figure 12.7 Placement of geosynthetic in reinforced slopes: (a) uniform spacing, (b) closer spacing where stress is higher, (c) secondary reinforcement placed to make compaction at the slope more effective, and (d) variable length, with longer primary reinforcement placed where stress is higher.

12.6.2.6 Designing embankment slopes with geotextiles

Steep fill slopes can be reinforced by multiple layers of geosynthetics, Figure 12.7, in various arrangements. A number can be envisaged where the reinforcement spacing is less for areas of higher stress. Shorter lengths of geosynthetics can be added to improve compaction of the soil near the embankment edge. The primary sheets of geosynthetics must be long enough to extend past the failure surface, and to have enough anchorage. If the strength and stiffness of the geosynthetic are directional, it is important to have the geosynthetic placed with the stronger direction in the direction of the major horizontal stress.

The design problem is one of multiple applications of the analysis for embankments on geosynthetics, Section 12.6.2.5. Equations 12.15 and 12.16 result from the development of Figure 12.8 [12.2].

$$FS = \frac{\sum_{i=1}^{n}(N_i \tan\phi + c\Delta l_i)R + \sum_{j=1}^{m} T_j y_j}{\sum_{i=1}^{n}(w_i \sin\theta_i)R} \tag{12.15}$$

$$FS = \frac{\sum_{i=1}^{n}(N_i' \tan\phi' + c'\Delta l_i)R + \sum_{i=1}^{m} T_j y_j}{\sum_{i=1}^{n}(w_i \sin\theta_i)R} \tag{12.16}$$

where

FS is the factor of safety, which should be >1.3 (unitless)

$N_i = w_i \cos\theta_i$, where w_i is the weight of the ith slice, and θ_i is the angle between the bottom face of the slice at its centerline and horizontal (kN/m or lb/ft)

Δl_i is the arc length of the *i*th slice (m or ft)
R is the radius of the failure circle (m or ft)
ϕ, ϕ' are the total and effective angles of friction
c, c' are the total and effective cohesion (kN/m^2 or lb/ft^2)
T_j is the allowable geotextile tensile strength of the *j*th layer (kN/m or lb/ft)

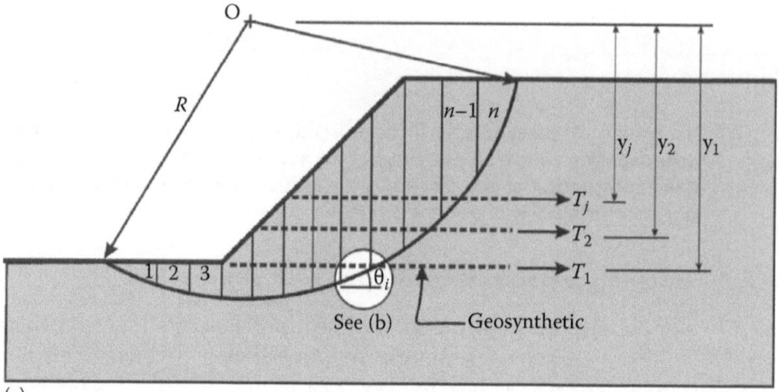

(a)

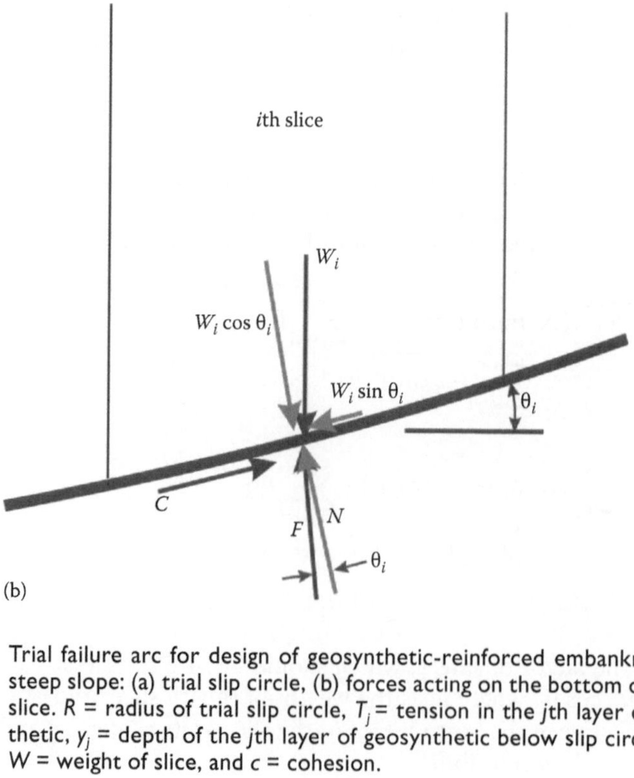

(b)

Figure 12.8 Trial failure arc for design of geosynthetic-reinforced embankment with steep slope: (a) trial slip circle, (b) forces acting on the bottom of a typical slice. R = radius of trial slip circle, T_j = tension in the *j*th layer of geosynthetic, y_j = depth of the *j*th layer of geosynthetic below slip circle center, W = weight of slice, and c = cohesion.

y_j is the moment arm for the *j*th layer of geosynthetic (m or ft)
n is the number of slices (unitless)
m is the number of layers of geosynthetic (unitless)
$N'_i = N_i - u_i \Delta x_i$ and $u_i = h_i \gamma_i$ the pore water pressure, where h_i = the height of the water level above the slip circle for the *i*th slice, Δx_i = the width of the slice and γ_w = the unit weight of water (kN/m or lb/ft)

Again, the geotextile layers should be checked for adequate anchorage (recall Equation 12.12).

12.6.2.7 Designing for subgrade stabilization with geotextiles

The functions the geotextile is expected to fulfill where an unpaved granular road is to be built on a soft subgrade are separation and filtration. The geotextile may also reinforce the road structure, but it will need to deform significantly in order to mobilize resistance. The road would need to tolerate substantial rutting or the cross section would have to be rutted during construction and then filled in.

If circumstances are right, the deformed geotextile will provide additional vertical resistance to the applied load with the vertical component of the tension in the fabric. Either the road structure will be stronger, or a thinner base can be used.

There are a number of design aids. The simplest approach is to apply a reduction factor to the base thickness to reflect the contribution the geotextile makes to the strength of the road structure. The base thickness is calculated using one or another design method, and then it is reduced by the reduction factor (which is <1).

Many manufacturers provide free software to calculate the unreinforced and reinforced road thickness. The software is particular to the manufacturer's geotextiles, and in some cases has tight restrictions on the range of input variables, but the software packages can be extremely useful.

Giroud and Noiray [12.9] developed a method based on the pressure exerted on the subgrade with and without a geotextile at the base/subgrade interface, and the bearing capacity of the subgrade. With no geotextile, Equation 12.17 can be derived:

$$c = \frac{P}{2\pi\left(\sqrt{(P/p_c)} + 2h_o \tan\alpha_o\right)\left(\sqrt{(P/2p_c)} + 2h_o \tan\alpha_o\right)} \tag{12.17}$$

where

c is the soil's cohesion (kN/m^2 or lb/ft^2)
P is the *axle* load (kN or lb)
p_c is the contact pressure between tire and road, taken to be the tire inflation pressure (kN/m^2 or lb/ft^2)

h_o is the aggregate layer thickness (m or ft)
α_o is the angle with vertical distribution of stress through the aggregate layer (26°)

A companion equation is derived, for the case where a geotextile is used, and it deforms enough by rutting that it can contribute to the resistance to the wheel load, Equation 12.18:

$$(\pi + 2)c = \frac{P}{2(B+2h\tan\alpha)(L+2h\tan\alpha)} + \frac{E\varepsilon}{\alpha\sqrt{1+(\alpha/2S)^2}} \tag{12.18}$$

where
c is the soil's cohesion (kN/m² or lb/ft²)
P is the *axle* load (kN or lb)
B and L are the width and length of the assumed contact patch where the tire contacts the road surface (m or ft)
h is the aggregate layer thickness (m or ft)
α is the angle with vertical distribution of stress through the aggregate layer

The two equations can be plotted for use as design aids, Figure 12.9.

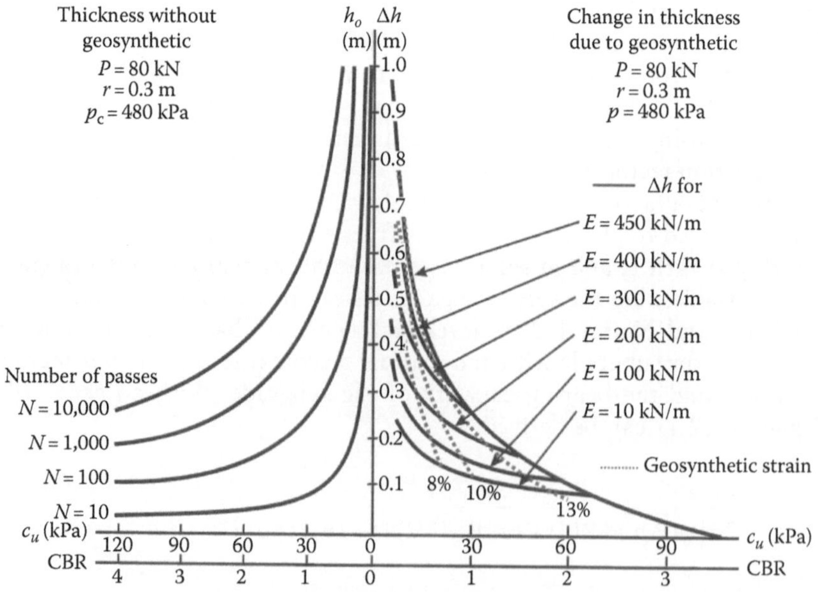

Figure 12.9 Design chart for geosynthetic-reinforced unpaved roads on soft subgrades. (After Giroud, J.P. and Noiray, L., *J. Geotechn. Eng. Div., ASCE*, 107(GT9), 1233, 1981; CGS, *Canadian Foundation Engineering Manual*, 4th edn., Canadian Geotechnical Society (CGS) c/o BiTech Publisher Ltd., Richmond, British Columbia, Canada, 488pp., 2006.)

The difference in thicknesses calculated in the two equations is the amount of aggregate saved by using a geotextile at the road base/subgrade interface. The cost saved is compared to the cost of supplying and installing the geotextile, and depending on the results, the geotextile is or is not installed. Even if it is not economical to install the geotextile for reinforcement, there may still be a need for a geotextile to provide separation and filtration. If a geotextile is used, beyond the requirements for base reinforcement, it will also have to satisfy the survivability requirements.

Geotextiles are not used to reinforce paved roads because there isn't sufficient deformation of the geotextile for it to mobilize any resistance to the loading. There may still be a need for the geotextile for separation and filtration, however. Again, the survivability requirements also must be met.

12.6.2.8 Designing for filtration with geotextiles

Designs for filtration are required, for example, for French drains along the edges of pavements. A trench excavated along the pavement edge is lined with a suitable geotextile, filled with a free-draining material, the tops of the geotextile folded closed, and the filled trench covered (Figure 12.10).

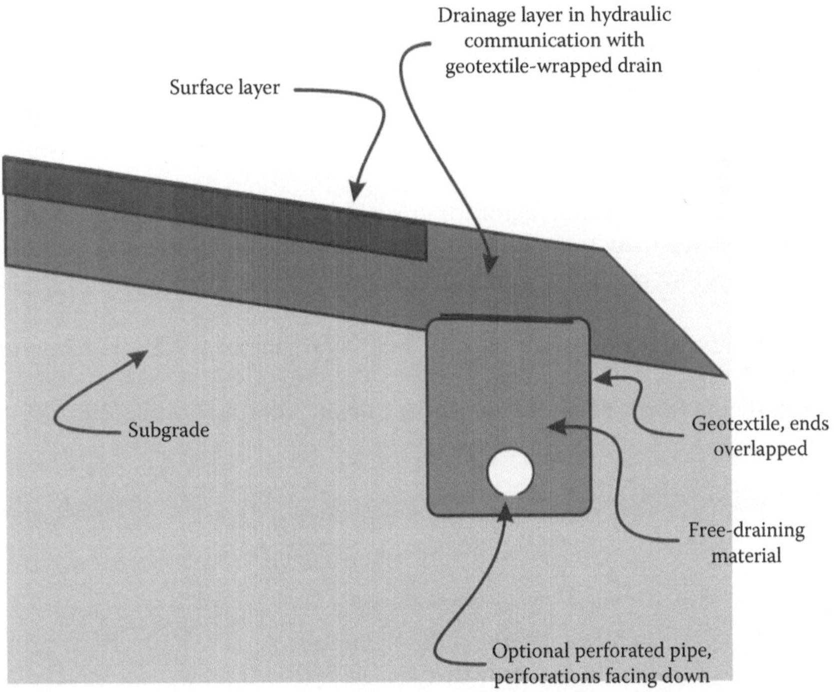

Figure 12.10 Design of geotextile-wrapped French road edge drain.

The trench may or may not have a perforated pipe installed near its bottom. The purpose of the geotextile is to filter the fines of the subgrade soil, to permit groundwater to enter the filled trench and drain away, while preventing the fines of the subgrade soil from entering the drain. The two filtration criteria—adequate permittivity and soil retention—must be satisfied.

In design, a trial and error approach is taken, where a particular geotextile is selected and its permittivity and soil retention checked for adequacy.

To check permittivity, the first step is to calculate the actual rate of flow into the geotextile, as determined by the project's requirements. This could be, for instance, the flow per metre length of a French drain. The required permittivity is then calculated from the actual flow rate, Equation 12.2. Next the allowable permittivity is computed from the selected geotextile's ultimate (i.e., manufacturer's lab test) permittivity, Equation 12.19:

$$\psi_{allow} = \psi_{ult}\left[\frac{1}{RF_{SCB} \times RF_{IN} \times RF_{CR} \times RF_{CC} \times RF_{BC}}\right] \tag{12.19}$$

where

ψ_{allow} is the allowable permittivity (s^{-1})
ψ_{ult} is the ultimate permittivity (s^{-1})
RF_{SCB} is the reduction factor for soil clogging and blinding (unitless)
RF_{IN} is the reduction factor for intrusion of soil particles into the geotextile (unitless)
RF_{CR} is the reduction factor for creep reduction of void space (unitless)
RF_{CC} is the reduction factor for chemical clogging (unitless)
RF_{BC} is the reduction factor for biological clogging (unitless)

Note that typical values for selected reduction factors [12.1] are given in Table 12.13.

Finally the factor of safety on flow is calculated, Equation 12.20:

$$FS_{flow} = \frac{\psi_{allow}}{\psi_{reqd}} \tag{12.20}$$

where

FS_{flow} is the factor of safety on flow (unitless)
ψ_{allow} is the allowable permittivity (s^{-1})
ψ_{reqd} is the required permittivity (s^{-1})

To perform as a filter, the geotextile must also retain the subgrade soil particles, and not let them wash into the free draining material. One of the retention criteria is selected (see [12.2, p. 101] for a suite of criteria), and the suitability of the geotextile determined.

12.6.2.9 Designing silt fences

Silt fences are used to retain the sediment liberated during a construction project, and prevent it from entering watercourses. Geotextile is strung between posts approximately a metre (3 ft) in height, across the natural drainage path. In some installations, wire mesh is attached to the posts to support the geotextile. The geotextile is keyed into the ground to prevent the sediment-laden water from short-circuiting under it (Figure 12.11).

While Koener [12.2] provides the details of a complicated model for designing silt fences, it is unlikely that one would design a silt fence from first principles. Manufacturers have marketed successful silt fence geotextiles for decades. Their advice should be sought directly.

An issue connected with the design of silt fences is their *maintenance*. Far too often silt fences are installed around a construction site in good faith, but they're soon forgotten, become tattered and torn, and cease to fulfill their function. They need to be inspected periodically, and maintained.

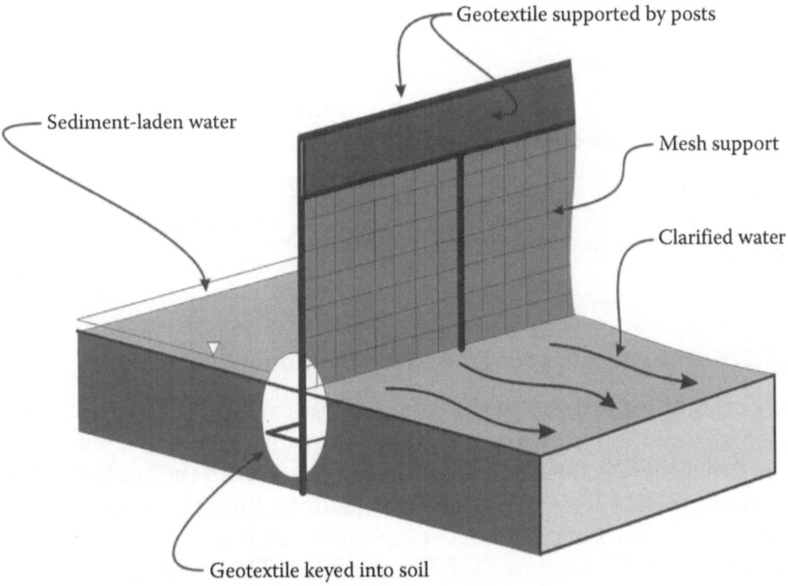

Figure 12.11 Design of silt fence, viewed from outside the construction area (water flows toward viewer).

12.6.3 Designing with geogrids

12.6.3.1 Allowable versus ultimate geogrid properties

Allowable and ultimate properties for geogrids are dealt with in the same way they are with geotextiles (recall Section 12.6.2.3 and Equation 12.5), with reduction factors reflecting the loss of strength caused by various individual factors, Equation 12.21:

$$T_{allow} = T_{ult}\left(\frac{1}{RF_{ID} \times RF_{CR} \times RF_{CD} \times RF_{BD}}\right) \tag{12.21}$$

where

T_{allow} is the allowable tensile strength (kN/m or lb/ft)
T_{ult} is the ultimate tensile strength (kN/m or lb/ft)
RF_{ID} is the reduction factor for installation damage (≥1) (unitless)
RF_{CR} is the reduction factor for creep (≥1) (unitless)
RF_{CD} is the reduction factor for chemical degradation (≥1) (unitless)
RF_{BD} is the reduction factor for biological degradation (≥1) (unitless)

Typical values for selected reduction factors [12.1] are given in Table 12.12.

12.6.3.2 Designing embankment foundations and slopes with geogrids

The approaches to designing embankment foundations and slopes with geogrids are the same as they are for geotextiles. Refer to Sections 12.6.2.5 and 12.6.2.6.

12.6.3.3 Designing for subgrade stabilization with geogrids

The situation for stabilizing the subgrade beneath and unpaved aggregate road with a geogrid is analogous to the case with a geotextile, with the difference that there is interlocking of the aggregate with the geogrid. Giroud and Han [12.10,12.11] have derived equations for this design case, and produced design charts. There is some controversy about the method's use of *J*, the geogrid's aperture stability modulus. This is the grid's in-plane torsional stiffness: a node is twisted in-plane, and the torque divided by the rotation.

In practice, the equations and charts presented by Giroud and Han can be used, or individual geogrid manufacturer's software used.

REFERENCES

12.1 Koerner, R.M. 2012. *Designing with Geosynthetics*, 6th edn., Vol. 1. Bloomington, IN: Xlibris.

12.2 Koerner, R.M. 1994. *Designing with Geosynthetics*, 3rd edn. Englewood Cliffs, NJ: Prentice-Hall Inc., 783pp.

12.3 McKean, E. (ed.). 2003. *Oxford American Dictionary and Thesaurus*. Oxford, U.K.: Oxford University Press, 1891pp.

12.4 CGS. 2006. *Canadian Foundation Engineering Manual*, 4th edn. Richmond, British Columbia, Canada: Canadian Geotechnical Society (CGS) c/o BiTech Publisher Ltd., 488pp.

12.5 Sarsby, R.W. 1985. The influence of aperture size/particle size on the efficiency of grid reinforcement. In *Proceedings Second Canadian Symposium on Geotextiles and Geomembranes*, Edmonton, Alberta, Canada, pp. 7–12.

12.6 AASHTO. 2006. *Standard Specification for Geotextile Specification for Highway Applications*, M288. Washington, DC: American Association for State Highway and Transportation Officials (AASHTO).

12.7 Rowe, R.K. and Soderman, K.L. 1985. Geotextile reinforcement of embankments on peat. *Geotextiles and Geomembranes* 2(4):277–298.

12.8 Rowe, R.K. and Soderman, K.L. 1986. Reinforced embankments on very poor foundations. *International Journal of Geotextiles and Geomembranes* 4(1):65–81.

12.9 Giroud, J.P. and Noiray, L. 1981. Geotextile-reinforced unpaved road design. *Journal of the Geotechnical Engineering Division*, ASCE 107(GT9):1233–1254.

12.10 Giroud, J.P. and Han, J. 2004. Design method for geogrid-reinforced unpaved roads, I, development of design method. *ASCE Journal of the Geotechnical and Geoenvironmental Engineering* 130(8):775–786.

12.11 Giroud, J.P. and Han, J. 2004. Design method for geogrid-reinforced unpaved roads, II, calibration and applications. *ASCE Journal of the Geotechnical and Geoenvironmental Engineering* 130(8):787–797.

Part III

Road construction and maintenance

Chapter 13

Road construction

13.1 INTRODUCTION

With the completion of the design work, a project moves to the construction phase. Construction needs to be planned and programmed, where programming is the timing of each step in the construction project. Budgets need to be established and tracked. If the project work is to be performed by a contractor (rather than being carried out by the road authority or road owner), contract documents need to be prepared, the project advertised and put out to tender, contractors' bids received and evaluated, and the project awarded to the successful bidder. Risks of delays and cost overruns need to be addressed. Equipment, material, and labor requirements need to be determined.

13.2 PLANNING

After the location of the road and its design are completed, the proposal accepted including the granting of environmental permits, funds allocated, and land obtained and surveyed, planning begins with the project being broken down into its components. Environmental issues that must be addressed include the following:

- Impacts on the landscape and esthetics
- Impacts on flora and fauna
- Impacts on surface water and groundwater
- Noise, vibration, dust, mud, sediment
- Sites of historical significance
- Sites of cultural significance
- Social issues
- Disruption of established traffic patterns

Permits covering many of these issues must be obtained from the relevant authorities. The time taken for the preparation of permit applications, submission of the applications, and winning their final approval can be significant.

Table 13.1 describes the project's components which must be planned and programmed.

13.3 BUDGETING AND PROGRAMMING

13.3.1 Estimating the budget

Estimating the cost of each component of the project in order to prepare a budget follows the same pattern. All of the activities in the component are listed. Unit rates and quantities are established for the labor, equipment, materials, and resources needed for each activity. The cost of each activity is then calculated as the unit rate times the quantity. Contingencies must be built into the budget to account for delays due to bad weather, supplier delays, changes in site conditions, contractor delays, and any other unforeseen conditions that can affect actual performance of the activity. As the project proceeds, the actual costs and quantities are compared to those that were estimated, and the budget is updated to account for the differences between the estimates and the actual costs.

13.3.2 Programming: Sequencing and timing the work

The estimated length of time to carry out each activity in the component, as well as the expected start time, is established. The order in which the activities must occur are set out. Bar charts help to visualize the programming of the activities (Figure 13.1).

Usually some of the activities in a component can occur simultaneously, saving time. In some cases activities must wait until one or more other activities are complete. The critical path method models the activities of a component as a network and the "critical path" is the sequence of activities that takes longer to complete than any other sequence of activities [13.1]. A diagram is prepared where each activity is represented by an arrow, running from left to right.* A node (circle) is placed at each end of the arrow. The left-hand node indicates the time at which the proceeding activity must be completed and at which the current activity can start. The node at the right-hand end indicates the time at which the current activity will end. A network diagram is built up showing the sequence and relationships of all the activities together with their starting and ending times (Figure 13.2).

* The length of the arrow is immaterial—it does *not*, for instance, represent the length of time taken to complete the activity.

Table 13.1 Project planning components

Component	*Items*
Preparation of the right of way	Clearing and grubbing
	Clearing buildings and debris
	Stripping
	Locating, protecting, and/or moving utility lines
Preparation of the subgrade (formation)	Earthworks: cuts and fills
	Diverting surface drainage
	Rock excavation
	Preparing materials
	Subgrade finishing including trimming and compaction
Establishing drainage	Surface drains
	Retention ponds
	Drainage ways
	Cross-drain culverts
Structures	Foundations for bridges and major culverts
	Bridge superstructures
	Major culverts
	Retaining walls
	High/steep slope stabilization
Pavement	Sub-base
	Base
	Surface layer
	Shoulders
	Abutting property connections
	Footpaths
Roadside features	Topsoil, seed, vegetation reestablishment
	Erosion protection
	Landscaping
	Rest areas, parking, landings, turnouts
Traffic facilities	Signs, signals, delineation, road markings
	Guiderails
	Lighting
Safety	Worksite safety, barriers, signs, fences
	Traffic control
	Public safety
Overheads	Project administration
	Contract administration
	Budgeting
	Programming
	Documentation
	Inspection
	Supervision, surveillance
	Commissioning and handover to owner

Source: Lay, M.G., *Handbook of Road Technology*, 4th edn., Spon Press, New York, 933pp., 2009.

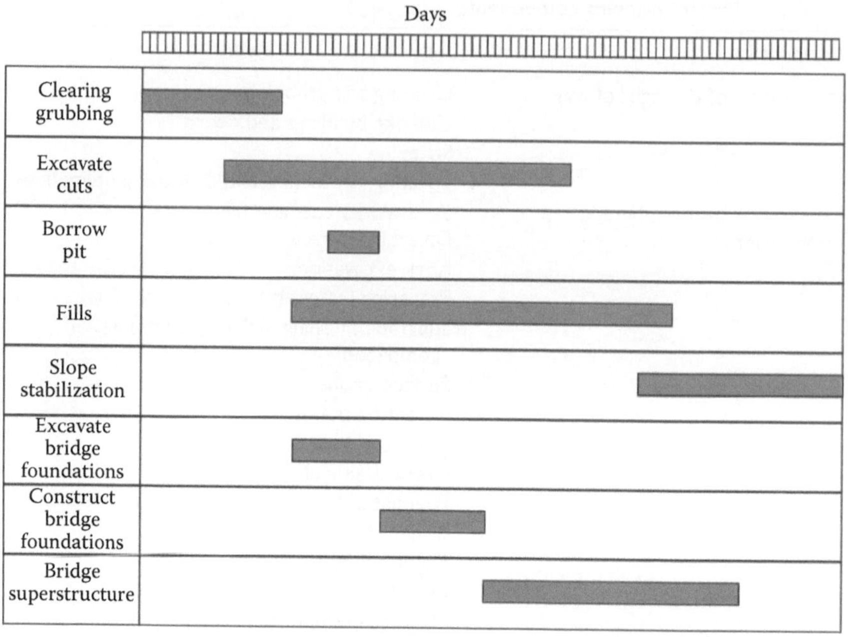

Figure 13.1 Simple example of a bar chart.

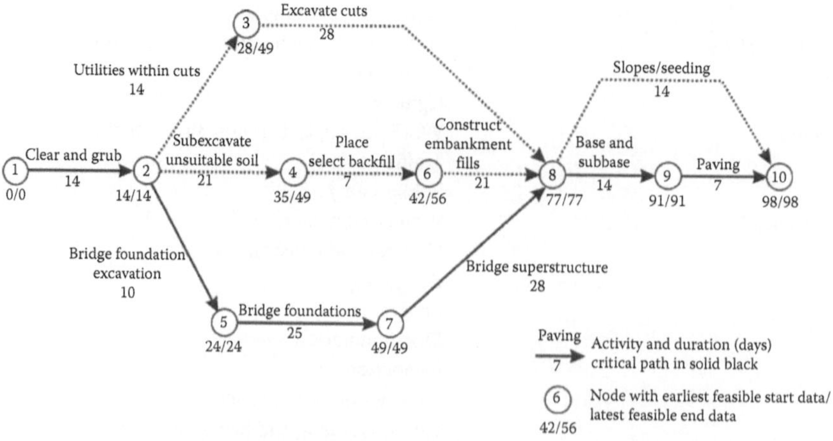

Figure 13.2 Simple example of a critical path network.

The diagram is invaluable in scheduling the acquisition of materials, timing the start of subcontracts and trades, and identifying bottlenecks. See Dykstra [13.2] for a full treatment of the method. Note that the method can be applied to the components within a project in the same way it is applied to the activities within a component.

13.4 CONTRACTUAL ARRANGEMENTS

Space does not permit going into great detail about contracts and contract law, but it is worthwhile to cover them briefly. Many texts dealing with contract law exist; much of what is presented here has been distilled from Barrie and Paulson [13.3].

13.4.1 Traditional arrangements

The traditional approach to civil design and construction as it began revolved around three distinct parties: the road owner,* the Engineer, and the contractor. The owner was the party needing the work constructed, and the contractor did the building. The engineer was responsible for the concept, design, and construction supervision. In Victorian times, the Engineer (note the capital "E"), could be a formidable personality, having full control of projects and their implementation (Figure 13.3).

The traditional arrangement continues, but it is by no means the only contractual arrangement in use. Still, a large number of contracts are arranged with the owner, engineer (or more generally, an engineering firm), and the contractor interacting as shown in Figure 13.4. The engineer and owner enter into a contract, under which the engineer provides design services, which may include environmental studies and permit acquisition, geometric design, subsurface and materials investigations, and possibly construction supervision services. With a second, separate contract, the owner hires a general contractor (or more generally, a contracting firm) to build the project. The general contractor in turn will likely hire specialist subcontractors to provide specific construction expertise, for example, blasting, provision, and construction of large culverts, or perhaps bridge erection.

The engineer in this arrangement is called upon to act as a referee at times, settling disputes between the owner and contractor. Clearly a great deal of professionalism is needed, especially given that one of the parties in such a dispute (the owner) is the party paying the engineer's fee. Other sensitive issues arise when the engineering firm doing the design for the project is not subsequently hired to provide construction supervision services. When this happens, the designer has no mechanism to ensure the

* In this discussion, the term "owner" will be used to denote the road agency, road authority, or private road owner, as the case may be.

Figure 13.3 A classic photograph of the great Victorian engineer, Isambard Kingdom Brunel (1806–1859). Brunel—designer, engineer, entrepreneur—completed many famous civil and marine projects. He is seen here standing in front of the chains of the *Great Eastern*, one of three ships he designed and built, the ship which laid the first successful Atlantic cable. (From Pannell, J.P.M., *An Illustrated History of Civil Engineering*, Thames and Hudson Ltd., London, U.K., 376pp., 1964.)

design basis and assumptions are respected. In the last couple of decades, following in the tracks of a much more litigious medical profession, professional engineers have found themselves involved in a spiraling number of lawsuits. A great deal has been written on the subject—see for example, Nasmith [13.5].

13.4.2 Design-build projects

The design and construction functions can be combined in one firm, in the *design-build* arrangement (Figure 13.5), also called the "turnkey" approach. The owner hires one firm to provide both functions, simplifying the situation. Construction may proceed much more quickly, because design and construction can be performed concurrently, in the "fast track" mode. On the other hand, there are fewer checks and balances in this arrangement.

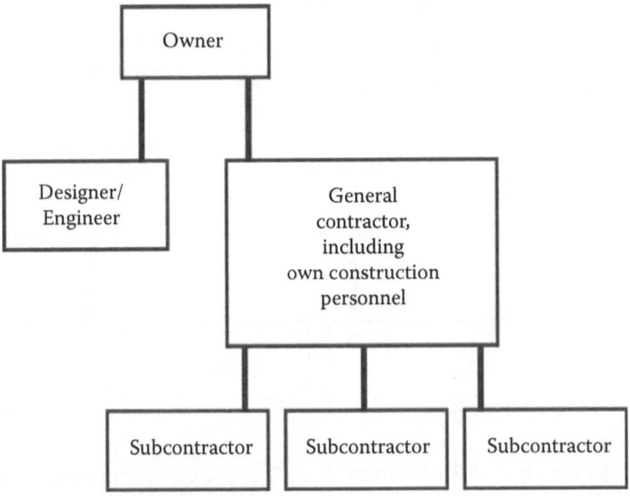

Figure 13.4 The traditional arrangement for construction contracts. (From Barrie, D.S. and Paulson, B.C. Jr., *Professional Construction Management*, McGraw-Hill Inc., Toronto, Ontario, Canada, 1992.)

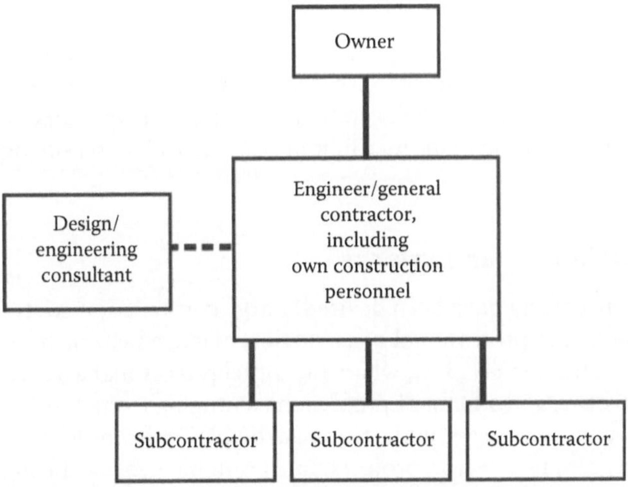

Figure 13.5 The design-build arrangement for construction contracts. (From Barrie, D.S. and Paulson, B.C. Jr., *Professional Construction Management*, McGraw-Hill Inc., Toronto, Ontario, Canada, 1992.)

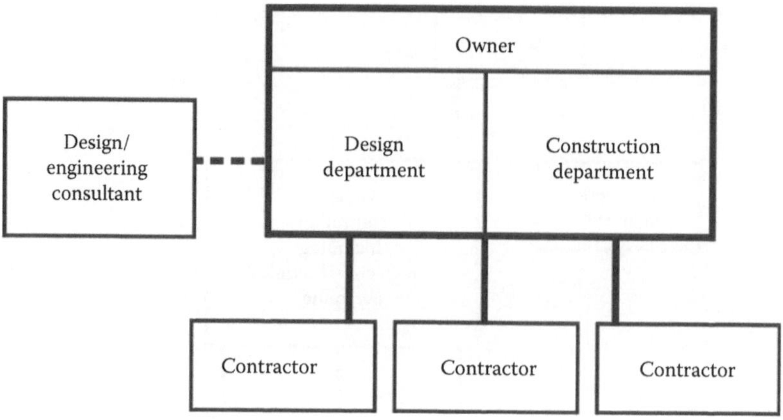

Figure 13.6 The owner–builder arrangement for construction contracts. (From Barrie, D.S. and Paulson, B.C. Jr., *Professional Construction Management*, McGraw-Hill Inc., Toronto, Ontario, Canada, 1992.)

It may also be that a specialist engineering and design consulting firm is hired, but in this arrangement, it is the general contractor, not the owner, who would enter into a contract with that firm. In addition, the general contractor may take on subcontractors to provide specialist construction expertise.

13.4.3 Owner–builder arrangements

A large owner, with a fairly steady construction program over many years, may gear up to accomplish the construction with its own forces (Figure 13.6). Design may be accomplished in-house, or it may be done by an external consulting firm. Again, it may be desirable to take on specialist contractors.

13.4.4 Other arrangements

Other arrangements have been devised. Large construction contracts can be carried out under professional construction management or program management arrangements [13.3], where high level project management skills are brought to bear, and a team of professionals ranging from designers through constructors is assembled, with strong participation from the owner.

Very large public sector projects can be implemented through design–build–operate arrangements, where a private company looks after all aspects of the project including financing, from the initial design through operation for a number of years, subsequently turning it over to the road owner at some price agreed at the outset. The facility must be brought to a stated condition before handover.

13.4.5 Checks and balances

In each of these arrangements, there are checks and balances on each of the parties involved in the project. In some cases, for example, the traditional arrangement, they are quite strong. The engineer is hired by the owner with the specific mandate to insure quality, cost, and speed of construction in the project. However, despite this strong control, it may well be that other arrangements are in fact better, for other reasons.

High quality work will be expected of a well-respected design-build firm, for example, and the speed of the construction, and thus the cost, may be much more favorable than they are in the traditional arrangement. Because they are implemented "under one roof," construction and the later stages of design can take place simultaneously. This parallel management of the work can dramatically speed up a project, and thus reduce its cost. But the freedom inherent in such an arrangement can be abused by less scrupulous firms.

How an owner elects to draw up its contracts thus is a matter of balancing checks and balances against the potential for speedier and less expensive construction. The wise owner makes it its business to know the relevant contractors and engineering firms, and their reputations. It should also be noted that there are of course "good" owners and "bad" owners. Good owners know how to get the best from their consultants and contractors, sustainably, over the long term, without making unreasonable demands of them.

13.5 FORMS OF CONTRACT

There are several contract forms, including lump sum, unit price, cost plus fixed fee, and cost plus incentive fee contracts. The first two are the most common.

13.5.1 Lump sum contracts

For a lump sum contract, the contractor quotes the final price for the complete project. Considerable thought goes into determining the bid price, and in the preparation of the bid, the contractor must work through applicable unit prices for time and materials, and give thought to what a reasonable profit will be. These must all be worked up knowing that some other competing contractor or contractors are engaged in the same process. The contractor wants to be competitive and win the job, but at the same time, must make a reasonable profit in order to stay in business.

Lump sum contracts are most suitable when the scope of the project is well known. The owner goes into the project knowing ahead of time almost certainly what the final cost will be, and so usually opts for a lump sum contract when there is a fixed budget which must not be exceeded.

13.5.2 Unit price contracts

The opposite is true of unit price contracts. If the scope of the work is *not* well known, a contract can be written with stated unit prices for the work of each of its elements. Such contracts contain costs per unit of time for labor and particular pieces of equipment to be used, and the costs of unit volumes or masses of materials to be used in the project. With this form of contract, an owner can begin a project of poorly known scope or conditions, and carry on either until the budget is exhausted or until the work is complete.

13.5.3 Role of the engineering firm

There are interactions between the form of contract and the contracting arrangement. The role of the engineering consultant, if involved, is different for each. If the arrangement is the traditional one and a lump sum contract form is used, the engineer is concerned with maintaining the *quality* of the work, as the incentive for the contractor is to go as quickly as possible, using the lowest acceptable quality materials.

On the other hand, if the traditional arrangement is used for a unit price contract, the engineer becomes time keeper and *quantity* checker. The incentive is for the contractor to go as *slowly* as tolerable, to use the most expensive materials, and to use as much of them as will be permitted.

It is extremely important that members of the engineering firm's construction supervision staff understand what their roles are, given the form of contract and the contracting arrangement.

13.5.4 Contracts in practice

In practice, most projects have elements of both forms of contract. Those parts of the work that have a well-known scope are best handled with lump sum bids, but there are other parts which are dealt with more appropriately with unit price quotations.

It is important to be alert to unbalanced bids. In such a bid, a contractor might quote a very low price for the lump sum part of the contract, where quantities are well known. This part will look particularly attractive to the owner; however, one or more of the unit prices may be abnormally high. This may not attract the owner's attention, since after all, the unit price items may or may not actually be required, or perhaps very little of them may be required. It may work out that the contractor in effect does the lump sum part of the work at little or no profit, yet reaps abnormally high profit on the high unit price items, resulting in a significantly higher overall project price (and contractor profit).

In one project, for instance, a powerful international company prided itself for having whittled down the bid price for a large earthworks project at one of its plants to nearly at-cost. The bulk of the work in the contract was performed on a lump sum basis, and indeed the contractor likely made no profit on that portion. However, as the project unfolded, there was the requirement for a surprising (to the owner) amount of associated unit price rock work, for which the contractor had quoted an abnormally high unit rate. In the end, because of the unusually costly rock work, the overall project cost was considerably higher than it would otherwise have been.

13.5.5 Contract bidding procedure

A glance at the business section of most newspapers will provide a glimpse into the construction bidding process. Notices are printed where various owners request the submission of bids for construction projects. The notices give details of the drawings and specifications that are available for perusal, may make mention of the possibility of inspection of the site, and state a date and time by which bids must be submitted by contractors interested in submitting a bid. The closing date and time are strictly enforced; if a contractor's bid is not received by the stated time, it will not be considered under any circumstances.

Usually there is also an indication of the policy on the selection of bids ("the lowest bid or any bid may not necessarily be selected"), and when notification of who the selected ("successful") contractor will be made.

Often the successful contractor is selected on the basis of weighted criteria. For example, a bid may be weighted 40% on cost, 20% on the stated understanding of the scope of the project, 20% on the proposed methods of construction, and 20% on the past experience with other relevant projects. In some cases, so cost considerations don't unduly influence the consideration of bids, recognizing that aspects beyond cost are important too, a "two envelope system" is used, where the cost bid is submitted in one envelope, and the rest of the bid details (proposed methods, personnel, relevant experience, etc.) in another envelope. The latter envelope is opened and evaluated first, followed by the opening and evaluation of the cost envelope. The weighted score is then calculated from the two evaluations, and the winning bid identified.

13.5.6 Changes, additions to the work, claims, entitlements

Very frequently in building construction, and usually to a lesser extent in earthworks, additional work, or changes to the scope of work will be identified by the owner or engineer after the construction has commenced. Arrangements are made for formal "work orders" in the former case, and "change orders" in the latter case.

It is possible that disagreement may arise as to whether work is indeed extra. The contractor may make a formal claim for additional payment for the work it deems extra. Formal processes are followed to decide the matter, but if it is unresolved, it becomes a dispute. An unresolved dispute can be settled through mediation, arbitration, litigation, or other means. The contractor is said to have an entitlement if the decision is that an additional amount should be paid beyond the amount in the original contract terms.

13.6 RISK

There is an element of risk in any construction project, and thus efforts are made to identify the risk and deal with potential exposures. Sources of risk include the following [13.3]:

- Delays in obtaining access
- Major design changes
- Strikes
- Failure of subcontractors
- Interruptions in supply of materials or equipment
- Unforeseen site conditions
- Weather problems
- Third-party interference

Some risks are best handled by the owner, others by the contractor, and still others by the engineer. Of course one strategy any of the parties could adopt is to shift the risk onto one of the other parties, and contract language must be carefully scrutinized by those other parties to be sure this does not happen. Shifting a risk to a party unable to control or manage it is unfair and is the source of considerable friction if permitted [13.3]. Improper assignment of risk can cause an unhealthy adversarial environment in contracts.

13.7 INSURANCE AND BONDS

Risk can be appropriately managed with fair assignment and proper contract language. In addition, various types of insurance and bonds are used to manage risk.

Types of insurance include worker's compensation insurance, comprehensive general liability insurance, contractual liability insurance, professional liability insurance, and builder's risk insurance.

Contractors are usually required to provide bid bonds, performance bonds, and payment bonds. Customarily amounting to 5%–20% of the total contract price, a bid bond is a guarantee that the contractor will indeed enter into a contract if declared the successful bidder. A performance bond of 100% of the contract price ensures that the owner will have sufficient funds to complete the project if the contractor defaults. Payment bonds provide funds for the owner to pay the subcontractors and suppliers should the general contractor default.

In addition to insurance and bonds, mechanic's lien legislation provides a means for a subcontractor, supplier, or other party contributing to the improvement of a property to obtain payment for their work. Until the work is declared complete, such a party has the right to claim payment, by recording a lien against the property.

13.8 PARTNERING

It has been recognized by owners, engineers, and contractors that an adversarial environment in construction contracting can lead to unsatisfying relationships and expensive litigation. In response, the Associated General Contractors of America have formalized a concept called "partnering" [13.6].

Partnering is intended to foster harmonious relationships on the job site, through the building of a team consisting of the owner, engineer, general contractor, subcontractors and suppliers, government agencies, and others with an interest in the project. In a formally partnered construction project, these stakeholders focus on solving problems, together, rather than adopting postures which eventually lead to time in court. The process is started after the awarding of a contract, but before construction begins, at a partnering session facilitated by a knowledgeable person with no connection to the project itself.

At the inaugural session, the stakeholders meet and get to know each other. The facilitator may lead them through some group dynamics exercises, which are useful both as ice-breakers and as training in how the members of groups can interact with one another to everyone's benefit. The partnering session usually works toward the development of a project-specific charter or mission statement composed by the members of the group, a scheme for resolving problems, and a list of problems to be tackled immediately.

The charter for one highway construction project with which the author was involved as facilitator read as follows:

> We the stakeholders are committed to work in a safe manner, as a team, in accordance with the contract, in a spirit of cooperation, professionalism, and integrity. Our goal is to achieve a quality project on time, while maximizing benefits, and minimizing negative impacts, to all parties.

The charter was signed by every participant in the inaugural partnering session. Each participant was provided a copy, and large versions were printed, to be hung in prominent locations in the site offices of the road authority and the contractor.

Stakeholders in the project included the governmental department of transportation; the general contractor; the subcontractors; the governmental department of fisheries; the governmental department of the environment; and the workplace health, safety, and compensation commission. In this particular project, no citizens' groups were involved, but they might be included in partnered projects. Other projects might involve other governmental agencies as well.

The problem resolution scheme devised during the partnering session for this particular project identified a number of steps, the personnel who would be involved at each step, and how long they would be given to resolve a problem. If the problem could not be resolved at the level of the personnel who first identified it, it would be moved to the next level, to be addressed by the personnel working at that next level. Table 13.2 lays out the scheme.

In the spirit of cooperation and partnering, the intention was that a problem would be addressed and solved at the lowest appropriate level, within the agreed time limit, or passed to the next level as soon as it became apparent that it could not be solved at the current level. Waiting for the time limit would just waste time.

During the partnering session, the team also identified 11 problems for which work on solutions could be started immediately. The problems ranged from identifying all government agency personnel for the contractor (four agencies were involved in the project) through to working up an environmentally satisfying solution for a particularly tricky blasting problem.

Past experience has shown that a sincere partnering effort does indeed result in more harmonious relations, far less litigation, and therefore less cost. However, to be successful, the concept must be openly, sincerely embraced by all senior personnel, and the commitment must remain strong throughout the project. Periodic top-up meetings of the stakeholders, focused on the partnering process, are required, beyond the regular technical project meetings. It is all too easy for individuals to revert to more adversarial behavior.

13.9 CONTRACT SPECIFICATIONS

The package of documents which is provided to bidders on construction projects includes a large number of drawings and also the specifications for the project. The specifications are the prescriptions for how the work must be done, what materials and/or resources are to be used, and the basis of payment for each item of work in the construction project. Government agencies such as departments of transportation spend prodigious effort

Table 13.2 Example problem resolution scheme

		Personnel involved in attempting resolution			
	Level	*For DoT*[a]	*For the contractor*	*For DFO,*[b] *DoE,*[c] *WHSCC,*[d] *respectively*	*Time limit to resolve problem*
Problem escalates ↑	5	Executive Director	Project Director	Director of Habitat Management Branch Director of Environmental Planning Branch WHSCC Appeals Tribunal	n/a
	4	District Engineer	Project Manager	Director of Habitat Management Branch Watercourse Alterations Program Manager Chief Compliance Officer	5 days
	3	Senior Resident Engineer	Project Manager	Environmental Assessment Officer Watercourse Alterations Technician Health and Safety Manager	2 days
	2	Resident Engineer	Project Engineer	Area Habitat Coordinator Regional Inspector Safety Officer	1 day
	1	Inspector or Technician	Project Engineer	Local Fisheries Officer Regional Inspector Safety Officer	1/2 day

The problem is addressed by the personnel who first identify it, at their level. If they cannot resolve the problem, or if their maximum time limit has been reached, they pass it to the personnel at the next level, and so on, up the levels until a solution has been devised.

[a] Department of Transportation.
[b] Department of Fisheries and Oceans.
[c] Department of the Environment.
[d] Workplace Health, Safety, and Compensation Commission.

in working up general specification documents, which are subsequently widely circulated among contractors and consultants. Additional project-specific specifications are written for each project. Individual owners in the private sector must either write their own specifications or hire knowledgeable consultants to produce them.

The format for specifications varies from owner to owner but the format is identical for every item in the set of specifications. For example, the typical specification for an item will likely include the following:

- *Description*: a short summary of what the specification for the item covers
- *Materials*: a list of the materials required for the item
- *Submittals*: a description of which forms, notifications, letters, etc., are required under for the item
- *Construction*: the details of how the relevant work is to be performed by the contractor (the meat of the specification)
- *Measurement for payment*: an indication as to how the item will be paid for, for instance, by the mass in tonnes, or the volume in cubic metres, or the running metre
- *Basis for payment*: unit price or lump sum, as appropriate

13.10 PAYMENT CERTIFICATES

In many earthwork contracts, construction takes place over a number of months or years. Of course the contractor finds it desirable to be paid as work progresses, so customarily payment is made periodically, usually monthly.

With the traditional contract arrangement, the owner, engineer, and contractor meet monthly to agree upon how much work has been completed since the previous monthly meeting, and what its value is. A payment certificate is drawn up, and the contractor is paid for that month's work—less a holdback of approximately 10%–15% or more of the value of the work.

The holdback is accumulated until the end of the project, as a surety that the work will indeed be completed. With the agreement and formal declaration of completion, the holdback is paid to the contractor.

How the progress of the work is tracked will depend on the specification. For earthworks, this may be a matter of measuring the mass of fill excavated, moved, placed, and compacted. Mass is checked with truck weigh scales or can be estimated from truck box or scraper volume measurements and agreed material densities.

On the other hand, the volume of fill placed can be measured *in situ*, using various surveying methods. Traditionally, stations are established at regular intervals along the road alignment, for instance, every 15, 20, 30, or 60 m

(50, 100, or 200 ft). Embankment cross sections are taken at each station on a periodic basis, and the cross-sectional areas at the stations used to determine the incremental volume of fill placed during that time period (e.g., month).

The two common methods of volume calculation are the average end area formula and the prismoidal formula.

13.11 EARTHWORK VOLUME CALCULATIONS

13.11.1 Average end area formula

In the average end area formula approach (Figure 13.7) to earthwork volume calculation, the cross-sectional area of the embankment fill is determined at each station, and the volume between stations calculated from Equation 13.1:

$$V = \frac{1}{2}(A_1 + A_2)L \tag{13.1}$$

where

V is the volume between stations (m^3 or ft^3)
A_1 is the cross-sectional area at the beginning station (m^2 or ft^2)
A_2 is the cross-sectional area at the end station (m^2 or ft^2)
L is the length between stations

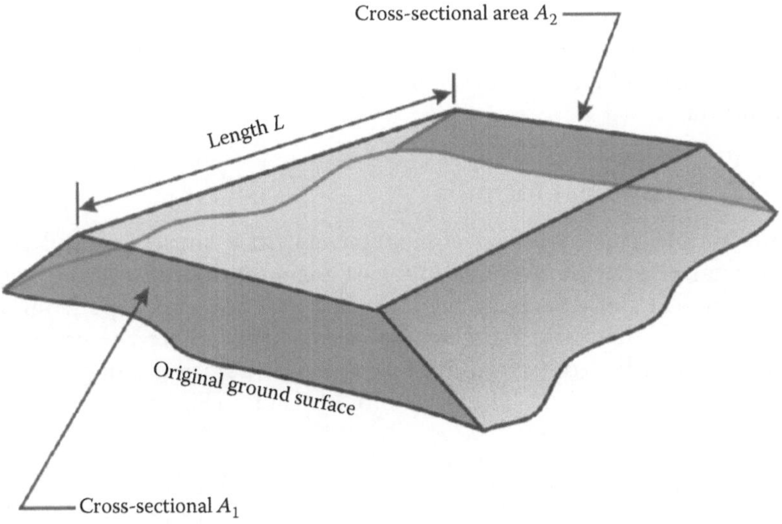

Figure 13.7 Earthwork volume determination using the average end area formula.

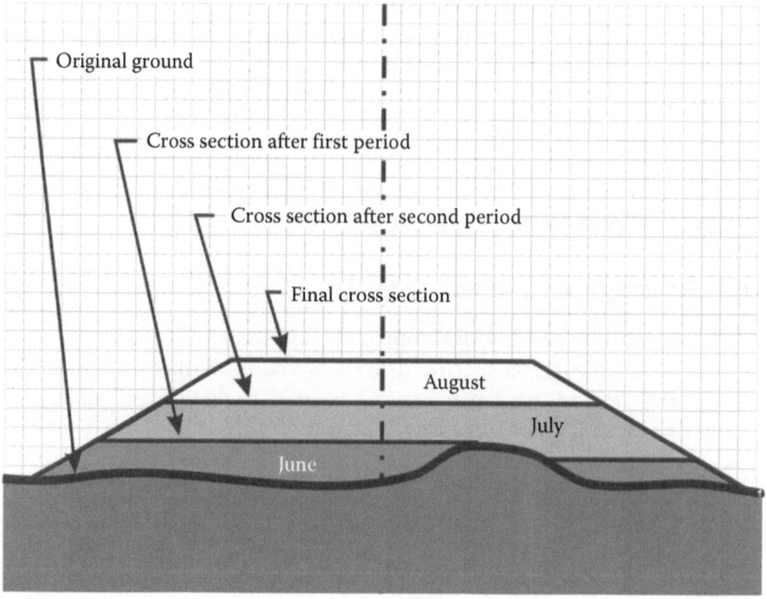

Figure 13.8 Example embankment cross sections at a particular station, plotted once each payment period.

The cross sectional surface profile is taken at each station each time period and plotted either manually or through a computerized approach, in order to determine the difference between the current cross-sectional area and the area determined in the previous time period's survey (Figure 13.8). Those areas are used in Equation 13.1 to find the volume of fill moved during that time period.

13.11.2 Prismoidal formula

While simple and requiring comparatively little surveying and computational effort, the average end area formula is progressively more inaccurate the more severely the cross sections change shape from one station to another. The most severe case occurs where the cross section is in full fill at one station, and by the next station along the road alignment, is in full cut. The prismoidal formula (Figure 13.9) is more accurate [13.7]:

$$V = \frac{1}{6}(A_1 + 4M + A_2)L \tag{13.2}$$

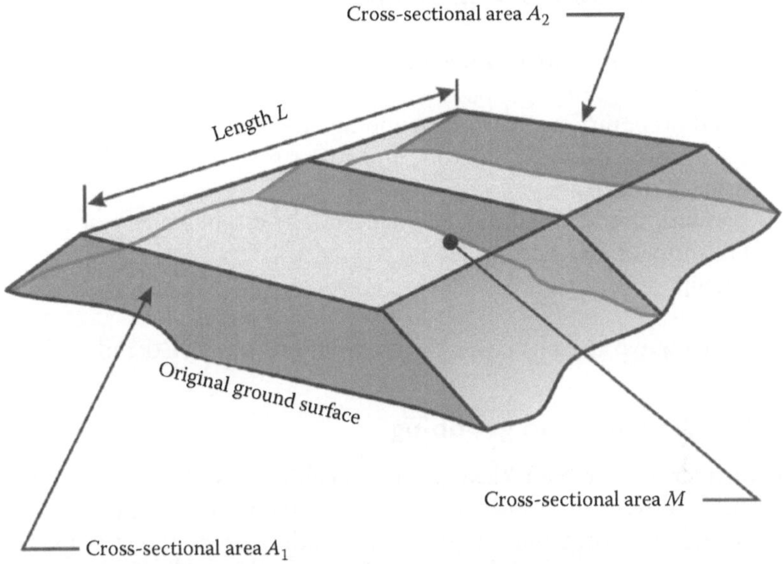

Figure 13.9 Earthwork volume determination using the prismoidal formula.

where

V is the volume between stations (m^3 or ft^3)
A_1 is the cross-sectional area at the beginning station (m^2 or ft^2)
A_2 is the cross-sectional area at the end station (m^2 or ft^2)
M is the cross-sectional area at a position midway between A_1 and A_2 (m^2 or ft^2)
L is the length between stations

It can be seen that the prismoidal formula involves a weighted average of the three cross-sectional areas, with the middle area getting four times the weight of the end areas. While the prismoidal formula is more accurate than the average end area formula, its use requires twice the surveying work, and twice the computational effort, for the same station interval.

13.11.3 Other methods

Any surveying method which facilitates the calculation of earthwork volume can conceivably be used. For instance, spot elevations across the fill might be taken periodically, and the data used to determine incremental volume. Such a method is well suited to the use of total station or GPS (global positioning system) survey equipment.

13.12 CONSTRUCTION SEQUENCE

Low-volume road construction generally consists of the following:

- Land preparation: clearing, grubbing, and stripping
- Excavation: roadway, borrow, and structural, including the establishment of drainage
- Formation of embankments, including the establishment of drainage
- Finishing operations
- Paving

The following paragraphs cover these operations in more detail.

13.12.1 Clearing and grubbing

The distinction between clearing and grubbing is that clearing is the removal of aboveground trees, shrubs, and vegetation, whereas grubbing is the removal of belowground vegetation including roots and stumps. Large or significant trees may be left intact and accommodated in an altered alignment of the road. Some of the trees to be removed may be merchantable, their careful removal and the sale of their logs contributing to the finances of the project. Buildings and other debris made to be disposed of by burning, if possible and permitted, or by burial, if possible and permitted.

13.12.2 Stripping

Stripping is the removal of topsoil and/or organic soil to expose mineral soil. Topsoil is a diminishing resource—it should be segregated, stockpiled, and protected for future use on slopes and for landscaping within the footprint of the project or outside.

13.12.3 Excavation

Some contracts distinguish between rock excavation and soil excavation because of the differences in the cost of them [13.1]. However, this can lead to ambiguity—"is it hard soil or soft rock?" Instead those contracts refer to "common excavation" rather than distinguish between the two types and provide the contractor with geotechnical information to help in the estimation of costs.

13.12.3.1 Roadway and drainage excavation

Roadway excavation begins with the grading and excavation of ditches to divert surface flows and to establish the new surface water drainage patterns. Approved excavated material won from cuts is stockpiled for later use in fills (or used immediately in filling operations). Material unsuitable for subsequent use in fills is deposited in dumps on site or wasted off-site in approved areas. The contract drawings show horizontal and vertical alignments, and cross sections—the roadway is to be built to these lines, and maintained at these elevations. Fill required to make up for settlement of the road embankment over time during construction must be accounted for. The roadway is to be maintained and kept free draining until paved.

13.12.3.2 Borrow excavation*

If sufficient material is not available in cuts to be used for the construction of roadway embankment fills, "borrow pits" must be established on site or nearby. With the grading of their floors, borrow pits must be maintained in a free-draining condition, to avoid the waste of time and material removing soil softened by rain or melting snow. Control measures must be implemented to deal with sediment generated by the exposed soil in the borrow pit. Borrow pits must be maintained in a safe condition for workers and the general public; steep slopes need to be protected, overhanging ledges removed, and safety fencing and notices erected. A plan for the closeout and rehabilitation of the borrow pit once construction is complete must be developed and implemented.

13.12.3.3 Excavation for structures

Excavations may be required for the construction of foundations; for the beds of large culverts; or for drainage structures such as sewers, catch basins, or access structures. These can be quite deep and time consuming to complete and may be on the critical path (Section 13.3.2), so must be started early in the construction sequence.

13.12.4 Embankment construction

Often the alignment of a low-volume road requires the construction of embankment fills. Fills are built from thin layers ("lifts") of material spread and compacted. Fills may also be needed to account for high groundwater conditions, soft subgrade soils, or permafrost. Their height will be dictated by the need to be above the water level and saturated subgrade soils, provide enough strength to counter weak subgrades, or to provide a thermal blanket above permafrost, respectively. Calculation is needed to determine their height, based on capillary rise, bearing capacity, or thermal modeling, respectively.

* "Borrow"—although this the traditional term, it is a curious word; it is highly doubtful that the material will ever be brought back!

Slopes must be stable in both the short term and long term. The slopes for embankments and cuts less than about 4.5 m high can be constructed with standard slope angles known to be stable for their material. For embankment and cut slopes with heights greater than about 4.5 m, slope stability analysis should be carried out. Embankment slope angles may be controlled by vehicle safety requirements; above a legally fixed height, the construction of barriers or flat slopes may be dictated.

To keep the working face dry with surface water draining away, embankment fills should be worked in the downhill direction and cuts should be worked in the uphill direction, wherever possible.

13.12.5 Pavement construction

The granular subbase and base layers of a pavement are conventionally laid in thin lifts, shaped with a crossfall for drainage (see Sections 8.2 and 8.3) and compacted to the required density. Care must be taken not to overwork the materials, as segregation may result—the coarser particles being separated from the finer particles, resulting in pockets of coarse particles that hold water and eventually generate potholes.

The final bituminous or cementitious layer, if they are to be built, is placed on the compacted base. Methods are discussed in the following references:

- *Chip-sealed surfaces*: Chipsealing in New Zealand [13.8]
- *Asphalt surfaces*: The Asphalt Handbook [13.9]
- *Portland cement concrete surfaces*: Portland Cement Association [13.10]

13.13 EQUIPMENT

Given the wide variety of types of equipment, the sizes the models are provided in, and the number of manufacturers, the array of equipment available for low-volume road construction can be bewildering. Table 13.3 lists the more common types of equipment and provides descriptions and remarks on their usage.

13.13.1 Equipment used to construct cuts and fills

Each piece of equipment used in earthwork cutting and filling operations has a maximum efficient earth movement distance, for example,

- Backhoes, excavators, shovels, loaders—very short distances
- Backhoes, shovels, loaders, bulldozers—short distances

Table 13.3 Equipment types

	Machine	*Running gear*	*Remarks*
	Backhoe	Rubber tires	• Tractor mounted • Combines excavator bucket and shovel • Light duty digging and loading
	Excavator	Steel tracks	• Light to heavy digging and loading • Digging action is toward the machine • Works from the top of a bank when loading
	Shovel	Steel tracks	• Heavy quarry work • Digging action is away from the machine • Works from the floor of the pit or quarry
	Loader	Rubber tires (can also be track-mounted)	• Light to heavy loading • Works from floor of pit
	Tandem dump truck	Rubber tires	• On- or off-highway hauling • Suitable for short to long distances
	Articulated dump truck	Rubber tires	• Off-highway use • Suitable for short- to medium-distance hauling
	Off-highway dump truck	Rubber tires	• Off-highway use • Suitable for short to medium distance hauling
	Bulldozer	Steel tracks	• Used for spreading or short-distance pushing • Some models have tilting blades • Sometimes equipped with a ripper
	Scraper	Rubber tires	• Self-loading and spreading • Suitable for short- to medium-distance hauling
	Grader	Rubber tires	• Light spreading and trimming • Sometimes equipped with a ripper

- Tandem dump trucks, articulated trucks, off-highway dump trucks, scrapers—medium distances
- Tandem dump trucks—long distances

What exactly constitutes very short, short, medium, and long distances depends on conditions, such as the material being transported and the terrain over which the material is being transported.

Earthmoving equipment works in teams, with the numbers and types of loading, transporting, spreading, and compacting machines being matched for the most efficient operation. The loading function can be performed by backhoes, excavators, shovels, or loaders. The transportation function can be performed by tandem dump trucks, articulated trucks, or off-highway dump trucks. Finally, the spreading function can be performed by bulldozers or graders, followed up by compaction by bulldozers or any of the wide variety of purpose-built compactors and rollers. Scrapers are unique in that they are a single machine capable of self-loading, transporting, and spreading the fill material.

More information is available on the management of equipment and the assembly of efficient teams of equipment in Lay [13.11], Wright and Paquette [13.1], and O'Flaherty [13.12].

13.13.2 Paving equipment

Various special machines exist to place, screed, and compact paving materials, specific to the paving material [13.8–13.10].

13.14 CONSTRUCTION MONITORING AND CONTROL: MONITORING COMPACTION

Construction monitoring and control is essentially a matter of ensuring the constructor meets the specifications for the project. The correct materials and components must be used and they must be brought to a condition which ensures the performance inherent in the design. A large proportion of construction control work focuses on compaction.

13.14.1 Compaction

Compaction is achieved by controlling soil or aggregate density (see Sections 6.3.4 and 6.3.5). The successful compaction of the soil depends upon five factors:

- Moisture content at the time of compaction
- The gradation of the soil or aggregate to be compacted

- The confinement of the soil or aggregate being compacted
- The type of equipment doing the compaction
- The compactive energy

While standard soil mechanics texts (e.g., [13.13]) cover soil compaction in detail, a summary can be given here:

- There is an optimum moisture content (OMC) for compaction. At water contents below the OMC, there isn't enough "lubricant" (water) to easily move soil particles into a denser arrangement under compactive effort. At water contents above the OMC, too much "lubricant" is available and soil particles are able to move away from the compactive effort. The OMC is unique to each soil or aggregate.
- Well-graded soils or aggregates have a favorable range of particle sizes to be compacted into a dense state. Uniform soils act like masses of single-sized marbles and are difficult to compact.
- If the soil or aggregate is not confined under compactive effort, the particles cannot be arranged densely. Confinement is achieved by ensuring the previously placed lift is adequately compacted, and that the soil or aggregate is placed in thin lifts.
- Fine-grained, cohesive soils are better compacted by equipment which imparts a kneading action, such as sheep's foot and pad foot rollers, whereas smooth-drum vibratory equipment works better on granular, noncohesive soils.
- As expected, using a higher compactive effort results in greater compaction.

Field compaction of fills and granular pavement layers is usually specified as a required minimum percentage of the standard Proctor maximum dry density (SPMDD, Section 6.3.5). Often a range of acceptable moisture contents at the time of compaction is specified as well, for instance, OMC $\pm 2\%$. In addition, the maximum loose lift thickness is specified. A rule of thumb for general fill compaction is that the lift thickness should not exceed one fourth of the diameter of the compacting roller. For granular pavement layers, specified lift thicknesses rarely exceed 150 mm (6 in.).

Specifications written this way constitute "end-result specifications." Some road authorities instead specify the details of the *method* of compaction, including the lift thickness, the type of compaction equipment, and the number of passes of the equipment. While it is easy to confirm that the required method has been followed by the contractor, following the method or perhaps changes in conditions may result in insufficient compaction. The contractor will have fulfilled contract requirements, but the end result—the level of compaction—may not be satisfactory.

13.15 CONSTRUCTION IN LABOR-INTENSIVE ENVIRONMENTS

The forgoing discussion has assumed that construction is carried out in an equipment-intensive environment, where labor is comparatively expensive and/or difficult to obtain. In developing countries, labor can be inexpensive and plentiful, whereas capital is expensive and/or difficult to obtain, and using large pieces of equipment and/or large fleets of equipment, together with their adequate maintenance, not feasible. Labor-intensive methods are far more appropriate.

The basic requirements (e.g., the requirements for adequate materials, the achievement of adequate compaction) remain the same. However, how they are achieved successfully in labor-intensive circumstances becomes the challenge. Methods must be adapted or at times even completely rethought, to be appropriate, and to respect the conditions.

The reader's attention is drawn to the published and current web-based information, including the following:

- The African Community Access Programme (AFCAP): http://afcap.org/SitePages/Home.aspx (Visited March 30, 2015)
- Publications of the South African National Roads Agency: http://www.nra.co.za/live/content.php?Session_ID=e76e6a56d6e0cf5586f4df4163b44d32&Category_ID=148 (Visited March 30, 2015)
- The Overseas Road Notes: http://www.transport-links.org/transport_links/publications/publications_list.asp?fcriteria=Overseas+Road+Notes&stype=Category&Categoryid=2 (Visited March 30, 2015)
- Highway and Traffic Engineering in Developing Countries [13.14]

REFERENCES

13.1 Wright, P.H. and Paquette, R.J. 1987. *Highway Engineering*, 5th edn. New York: John Wiley & Sons, 717pp.

13.2 Dykstra, D.P. 1984. *Mathematical Programming for Natural Resource Management*. New York: McGraw-Hill Book Company, 318pp.

13.3 Barrie, D.S. and Paulson, B.C. Jr. 1992. *Professional Construction Management*. Toronto, Ontario, Canada: McGraw-Hill Inc.

13.4 Pannell, J.P.M. 1964. *An Illustrated History of Civil Engineering*. London, U.K.: Thames and Hudson Ltd., 376pp.

13.5 Nasmith, H.W. 1986. *Suit is a Four Letter Word*. Vancouver, British Columbia, Canada: BiTech Publishers Ltd., 170pp.

13.6 Associated General Contractors of America. 1991. Partnering, a concept for success. Washington, DC: Associated General Contractors of America.

13.7 Bouchard, H. and Moffitt, F.H. 1965. *Surveying*, 5th edn. Scranton, PA: International Textbook Company, 754pp.

13.8 TNZ. 2005. *Chipsealing in New Zealand.* Wellington, New Zealand: Transit New Zealand, 524pp. http://www.nzta.govt.nz/resources/chipsealing-new-zealand-manual/chipsealing-in-new-zealand.html, visited March 30, 2015.
13.9 AI. 2010. *The Asphalt Handbook*, 7th edn. Lexington, KY: The Asphalt Institute (AI)., 788pp.
13.10 Portland Cement Association Bookstore. http://members.cement.org/EBiz55/Bookstore/BookstoreHome.aspx, visited March 30, 2015.
13.11 Lay, M.G. 2009. *Handbook of Road Technology*, 4th edn. New York: Spon Press, 933pp.
13.12 O'Flaherty, C.A. (ed.). 2002. *Highways: The Location, Design, Construction, and Maintenance of Road Pavements*, 4th edn. Oxford, U.K.: Butterworth-Heinneman, 553pp.
13.13 Craig, R.F. 1997. *Soil Mechanics*, 6th edn. London, U.K.: E & FN Spon, 488pp.
13.14 Thagesen, B. (ed.). 1996. *Highway and Traffic Engineering in Developing Countries.* London, U.K.: E & FN Spon, 485pp.

Chapter 14

Road maintenance

14.1 INTRODUCTION

As noted in Chapter 1, agriculture, business, commerce, and communications all depend heavily on low-volume roads. The roads must be well maintained or these functions will suffer. Trips to schools, shops, and hospitals all depend on well-maintained roads [14.1]. In addition, an efficiently managed road network is a requirement for successful trading countries. Vast quantities of agricultural, mining, and manufacturing products are carried by trucks on roads, and the expectation of users is that the roads will be maintained to cope with the inevitable deterioration. Further, there is the expectation that any disruption to service will be minimal *during* maintenance operations.

Maintenance management has been called a matter of "doing the correct thing at the correct time in the correct place" [14.1].

14.2 MAINTENANCE ADMINISTRATION AND PLANNING

14.2.1 Administration

An agency's overall efforts to protect the capital investment in road infrastructure is distributed among reconstruction, rehabilitation (or remedial maintenance), and maintenance (or preventive maintenance) [14.2]. Reconstruction is a matter of completely replacing the existing pavement, rehabilitation is carried out to restore the existing road to its original condition, and maintenance is done to stop or retard—rather than reverse—deterioration. The typical distribution of effort between these three is presented in Table 14.1. The figures are approximate and highly dependent upon the individual circumstances of the agency. Note that roughly half the effort is devoted to pavements, with typically about a fifth of the effort put into road maintenance.

Road maintenance itself can be divided a number of ways.

Table 14.1 Distribution of a road authority's work to protect road assets

Item	*Range (%)*	*Typical (%)*
New facilities pavement	10–60	30
Reconstruction	5–40	20
Rehabilitation	0–20	10
Maintenance	10–50	20
Bridge maintenance	0–10	5
Traffic management	0–10	5
Organizational management	5–20	10

Source: Lay, M.G., *Handbook of Road Technology*, 4th edn., Spon Press, New York, 933pp., 2009.

Giummarra [14.3], in Australia, identifies on-demand and preventive maintenance. Applied when a defect arises, on-demand maintenance is reactive. Defects are predicted well in advance, and actions recommended to reduce or to eliminate them in preventive maintenance. The approach used depends on the importance of the road to the network. Usually on-demand maintenance results in higher user costs (and inconvenience), whereas preventive maintenance usually requires higher initial costs [14.3].

Giummarra [14.3] also divides maintenance into routine and periodic maintenance. For the former, permanent patrols perform such work as light grading or clearing blocked drains and culverts. For the latter, teams are assembled, for example, for regravelling, recompacting, and reshaping projects.

Lay [14.2], in the United Kingdom, considers three types of maintenance:

1. Periodic, programmed maintenance
2. Routine maintenance
3. Service restoration

Periodic, programmed maintenance is predictable and planned. It includes operations such as mowing grass, cleaning ditches, repainting lines, grading shoulders, sealing joints, and filling cracks in the pavement. Routine maintenance is not amenable to detailed planning but can be assessed in advance. It includes operations such as repairing potholes and patching road surfaces. It typically has a high benefit–cost ratio. Finally, service restoration deals with work such as ice and snow removal, and cleanup and repair of flood and wind damage.

Periodic and routine maintenance are usually done by patrols of workers with general skills, whereas service restoration requires workers with skills in that particular type of maintenance [14.2]. For the former, the patrols usually consist of 3 or 4 workers equipped with the required tools and a truck. A patrol will operate over a defined part of the road network, say 100 km (60 miles) of road. A supervisor is assigned to up to about 10 patrols. The patrols are in many ways self-managing, and "pride of patch" is a strong motivator.

Another way to organize the maintenance effort is to have "lengthmen" responsible for a particular stretch of road [14.2]. Lengthmen have the necessary hand tools but no vehicle and are usually supported by patrols.

O'Flaherty [14.1] classifies maintenance by the pavement defects involved:

- Whole carriageway (roadway) minor deterioration—potholes, crazing, surface fretting, loss of chips, and excess bitumen
- Whole carriageway major deterioration—cracking, loss of aggregate
- Loss of skidding resistance
- Wheel track rutting
    - In the foundation (subgrade)—requires reconstruction
    - In the surface course—dealt with using surface work, such as overlays or mill and replace
    - Should check other defects, cracking, for example
    - Surface irregularities (deformations)

There are other items requiring maintenance, including road signs, road markings, bridges, culverts, and drainage structures.

14.2.2 Planning

Maintenance planning is needed to ensure constant levels of service, standards, methods, and procedures [14.2]. Lay [14.2] has developed a chart of the flow of maintenance management decisions and communications (Figure 14.1), depicted as a loop.

The loop begins with management setting policy and establishing funding. Standards are then set to meet policy. A maintenance program is planned, emphasizing the annual program. The plan is checked to determine if it is within the established budget. If it is, the next step is the scheduling of the work. If the plan exceeds the budget, there is a need to reassess the funding allocation and confirm or alter it. With the work program within the funding allocation, the work is scheduled, distributed, and executed. The next step is to monitor and assess the effectiveness of the plan. Whether the road needs were met is evaluated. Next year's needs are then developed, and the budget needed estimated again, back at the beginning of the loop.

14.2.3 Maintenance activities

Maintenance activities range from repairing subgrade failures to spraying roadside weeds. Jordan et al. [14.4] cited by Lay [14.2] have laid out a comprehensive list of activities, grouped by theme, Table 14.2. Note that in Table 14.2, Items 110 Paved Carriageways/Pavement Structure and 210 Unpaved Carriageways/Existing Surface have been expanded to show all Task Definitions, as examples. All the others can be expanded similarly. See Lay [14.2] for the full treatment.

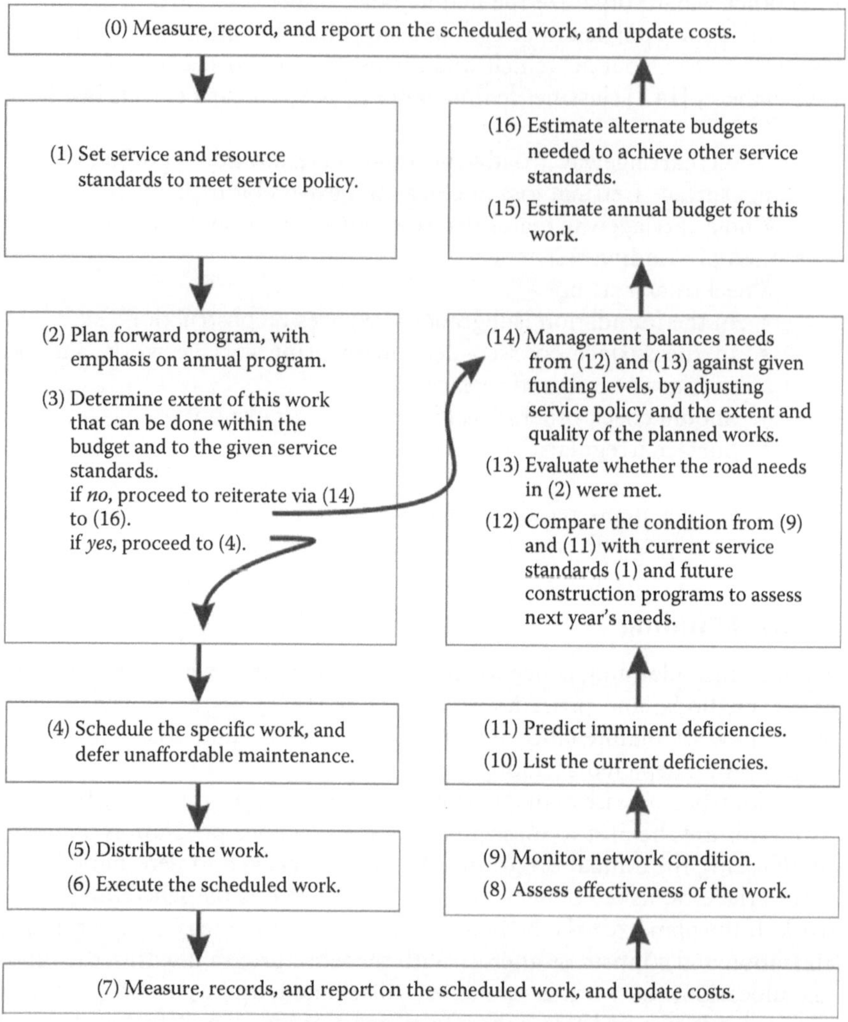

Figure 14.1 Maintenance management flow chart. (From Lay, M.G., *Handbook of Road Technology*, 4th edn., Spon Press, New York, 933pp., 2009; Jordan, J.R., Logue, G., and Anderson, G.M., Review of NAASRA study of road maintenance standards: Costing and management, *Proceedings of 10th ARRB Conference*, 10(2), 232 1980.)

14.3 CONCRETE PAVEMENTS

Concrete (rigid) pavements behave differently from bituminous (flexible) pavements, so their maintenance is different. Failures include spalling (pieces breaking off the surface) at joints, surface scaling, and cracking

Table 14.2 Road maintenance activities

Categories	*Task definition*
100	**Paved carriageway (roadway)**
110	*Pavement structure*
111	Repair localized subgrade failures
112	Repair localized pavement failures
113	Reconstruct short sections of up to 500 m^2 (5400 ft^2) in area
114	Clean joints in slabs and fill and seal cracks in pavements
115	Lift and fix displaced concrete slabs
120	*Pavement surface*
130	*Pavement edges*
140	*Unpaved shoulders*
150	*Paths*
160	*Access*
200	**Unpaved carriageways (roadways)**
210	*Existing surface*
211	*Grade to restore shape (dry maintenance)*
212	*Provide dust suppressants*
213	*Rip excess material*
214	*Remove loose material*
215	*Patch soft spots*
216	*Add new material*
217	*Compact with water (wet maintenance)*
218	*Spray weeds with herbicide*
220	*New surface*
300	**Drainage**
310	*Surface drains*
320	*Pits and sediment traps*
330	*Culverts and bridges with spans less than 6 m*
340	*Underground drains*
350	*Roofwater outlets in road reservation (right of way)*
400	**Roadside**
410	*Vegetation within road boundary (right of way)*
420	*Motorists' services*
430	*Litter control*
440	*Earthworks*
450	*Property/boundary fences and gates*
460	*Other service agencies within the road boundary (right of way)*
470	*Roadside hazards*
480	*Sites of environmental significance*

(Continued)

Table 14.2 (*Continued*) Road maintenance activities

Categories	*Task definition*
500	**Traffic facilities**
510	*Signs and markers*
520	*Road furniture*
530	*Road markings*
540	*Traffic signals*
550	*Roadside telephones*
560	*Lighting*
570	*Communications links*
600	**Structures**
610	*Bridge substructures*
620	*Waterways maintenance*
630	*Bridge superstructures*
640	*Bridge decks*
650	*Miscellaneous*
700	**Restoration**
710	*Traffic accidents*
720	*Storm and flood damage*
730	*Fire damage*
740	*Road openings due to utilities operating in the road reserve*
750	*Permanent reference points*
800	**Overheads**
810	*Administration of maintenance activities*
820	*Roadway inspection*
830	*Inspection and supervision of work*
840	*Depots and camps*

Source: From Lay, M.G., *Handbook of Road Technology*, 4th edn., Spon Press, New York, 933pp., 2009; Jordan, J.R., Logue, G., and Anderson, G.M., Review of NAASRA study of road maintenance standards: Costing and management, *Proceedings of 10th ARRB Conference*, 10(2), 232, 1980.

Note that for brevity, only Items 110 and 210 have been expanded as examples of showing the complete Task Definition. All other Task Definitions can be expanded similarly—see Lay [14.2] for the full treatment.

[14.1]. Because of their movement, jointed concrete slabs require periodic maintenance of the joints. Seals must be replaced because a loss of seal leads to water penetration, subsequent weakening, and deterioration of lower pavement layers. Base, subbase, and/or subgrade problems can sometimes be repaired by grouting. If this is not successful, it becomes necessary to remove the affected slab; repair the deteriorated base, subbase, or subgrade; and pour a new slab [14.1].

14.4 ASPHALT PAVEMENTS

14.4.1 Deterioration

Asphalt roads deteriorate in a number of characteristic ways, including the following [14.1]:

- Polished aggregate—the smooth texture of the particles leads to a reduction in skidding resistance
- Loss of surface texture—the road surface itself becomes smoother, also leading to a reduction in skidding resistance
- Deformation of the surface due to traffic—rutting and other deformation leads to water ponding on the road surface, contributing to the risk of hydroplaning, where a thin film of water builds between the pavement surface and the tread face of tires, resulting in a loss of vehicle control
- Cracking and surface deterioration caused by oxidation and embrittlement of the asphalt cement
- Fatigue strain of the foundation (subgrade) causing structural deterioration of the pavement

Maintenance treatments are decided once the type of deterioration is identified and understood.

14.4.2 Maintenance treatments

Maintenance treatments should be selected to take into account whole-life costs and should include user costs [14.1]. Treatments may be grouped as follows [14.1]:

- Surface dressings (treatments), thin surfacings
- Resurfacing
- Strengthening

Surface dressings are applied by spraying one or two layers of asphalt cement on the existing road surface, and then spreading and rolling-in single-sized chips which can be coated or uncoated. The treatment can be unpopular with users, when flying chips crack windshields and damage vehicle paintwork. Many other (often) proprietary thin surfacings can also be applied. They can restore skid resistance; treat narrow, stable cracks; improve impermeability; and restore the road's cross-sectional profile if not too bad [14.1].

Resurfacing represents a more substantial set of treatments than surface dressings. Approximately 40–50 mm (1.5–2.0 in.) of new asphalt is placed on the pavement. It may be placed directly on the road surface, or placement may come after a similar depth of the existing asphalt is milled and removed. The old material can be recycled in either hot-mix or cold-mix asphalt. The new surface may have better properties than the old surface had. For example, it may have better noise reduction, or improved skid resistance, or be less conducive to water spray from tires. Milling and replacing the old surface is usually more expensive than placing a simple overlay, but it may be that the rise in surface level ("grade raise") of the overlay cannot be tolerated at abutting pavements or curbs and gutters [14.1].

Strengthening pavements implies reconstruction or the placement of a structural overlay. It is usually done when the problem is more than surface deep. Suggested references for assistance with pavement and structural overlay design are shown in Table 14.3. As it is in most countries, UK practice is to design pavements to last typically 20 years, when a structural overlay is placed on the pavement [14.1]. Sufficient space is allowed under bridges and overhead signs to accommodate the future thickness of the pavement.

Ontario (Canadian) practice is to perform a pavement condition survey, observing and recording the severity and extent of 12 distresses. The distresses include two types of surface defects, three types of surface deformations, and seven types of cracking. The severity of a distress is how bad it is, and the extent is how much of the road segment exhibits the distress. The treatment of the pavement is then related to the dominant distress type, its severity, and its extent. Table 14.4 [14.5] presents the approach for the case where wheel path rutting is the dominant distress.

Table 14.3 Selected pavement design and overlay references

Country	*References*	*Remarks*
United States	[14.12] AASHTO 1993 *Guide for Design of Pavement Structures*	Guide is replaced by *Mechanistic Empirical Pavement Design Guide* (MEPDG) literature, but the "old" guide is in many ways more appropriate for LVR
Canada	[14.13] Adaptation of AASHTO Pavement Design Guide for Ontario Conditions	Interprets the AASHTO guide and provides regional inputs to pavement design routine
	[14.14] MTO Pavement Design and Rehabilitation Manual	Comprehensive treatment of materials, design, and rehabilitation—known as "the Blue Book"
United Kingdom	[14.15] TRRL Report LR833	Covers new design and overlays
Australia/New Zealand	[14.3] Unsealed Roads Manual together with the N.Z. Supplement	Focuses on only *unsealed* roads (gravel, earth-surfaced) but provides full treatment

Table 14.4 Example treatment list for a particular dominant distress—Rutting

Pavement condition evaluation		
Severity	*Density (extent)*	*Recommended treatments*
Very slight	Few	None
	Intermittent	
	Frequent	
	Extensive	
	Throughout	
Slight	Few	None
	Intermittent	
	Frequent	
	Extensive	
	Throughout	
Moderate	Few	None
	Intermittent	Machine chip seal Machine patching
	Frequent	Machine chip seal Machine patching
	Extensive	Machine chip seal Machine patching
	Throughout	Machine chip seal Machine patching Cold mill with hotmix resurfacing
Severe	Few	Machine patching
	Intermittent	Machine patching
	Frequent	Machine patching
	Extensive	Cold mill with hotmix resurfacing Hotmix resurfacing
	Throughout	Same alternatives as for "extensive"
Very severe	Few	Machine patching
	Intermittent	Machine patching
	Frequent	Machine patching
	Extensive	Rehabilitation candidate
	Throughout	Rehabilitation candidate

Source: Chong, G.J., Phang, W.A., and Wrong, G.A., *Manual for Condition Rating of Flexible Pavements*, SP-024, Ministry of Transportation of Ontario, Toronto, Ontario, Canada, 111pp., 1989 © Queen's Printer for Ontario, 2015. Reproduced with permission.

14.5 SURFACE-TREATED PAVEMENTS

Surface-treated pavements are also known as chip-sealed and tar and chip pavements. Because they are sealed, surface-treated pavements behave in a similar fashion to asphalt pavements. However, because the sealing layer is thin and depends on a well-built granular base for structural strength, surface-treated pavements also behave in some ways like gravel-surfaced roads. Following are discussions of the typical distresses exhibited by surface-treated pavements and the relevant maintenance treatments.

14.5.1 Deterioration

The deterioration of surface-treated pavements is manifested in distresses such as surface defects, surface deformation, and cracking. These can be broken down further as follows:

- Surface defects
  - Loss of cover aggregate
  - Streaking
  - Flushing
  - Potholes
  - Pavement edge breaks
- Surface deformation
  - Rippling
  - Wheel track rutting
  - Distortion
- Cracking
  - Longitudinal cracking
  - Transverse cracking
  - Pavement edge cracking
  - Alligator cracking

Chong et al. [14.6] provide detailed discussion of each of the distresses. A summary is given here.

Surface defects focus on potholes, edge breaks, and defects in the aggregate or asphalt cement component of the pavement sealing "system" (aggregate held in place by asphalt cement). Cover aggregate can be pulled off the surface under traffic, leaving the asphalt cement unprotected. During construction, alternating lines of lean and heavy asphalt cement may have been put down, parallel to the direction of traffic. If too much asphalt cement was placed during construction, or a new sealing layer is placed over an old one with excess asphalt cement, flushing may occur where free asphalt cement may migrate upward to the surface. If flushing occurs, it often happens in the wheel tracks. It is more common during hot weather.

As with gravel pavements, potholes can form where cross fall is lacking and water stands on the pavement, or if construction technique and quality control are poor, especially permitting aggregate segregation.

Frost action, poor edge support, poor drainage, or poor alignment where traffic runs close to the edge of the seal can lead to edge breaks.

Surface deformations can lead to poor riding quality. Rippling (washboarding) can occur when the pavement surface is pushed into regular undulations by traffic braking or accelerating. It is often accompanied by flushing. Wheel-track rutting can occur under repeated traffic load when the underlying granular layers deform. The base or subbase may be unstable due to excess pore water pressures. Other pavement distortions may occur as a result of settlement, frost action, lack of subgrade support, or embankment slope failure.

Cracking of a number of types can be identified. Longitudinal cracking can result from traffic loading, poor construction techniques, or environmental effects such as frost action or moisture changes. Transverse cracking can be caused by shrinkage in cold temperatures or frost action. Pavement edge cracking can happen if the pavement is subjected to frost action, has poor edge support, poor drainage, or traffic coming close to the edge.

Alligator cracking is seen when a surface-treated pavement is coming to the end of its structural life. Alligator cracking—cracks forming a network of polygonal blocks looking like the skin of an alligator—results from a softening of the pavement structure, normally caused by excess moisture. Alligator cracking is progressive.

14.5.2 Maintenance treatments

Maintenance treatments for surface-treated pavements usually seek to reestablish the seal, smooth the deformation, or address the cause of the bump or hole. Table 14.5 presents a summary of maintenance treatments for surface-treated pavements.

14.6 GRAVEL- AND EARTH-SURFACED ROADS

The condition of gravel* and earth-surfaced roads can change much more rapidly than the condition of sealed roads, and so they require regular maintenance at frequent intervals. An advantage of these unsealed roads, however, is that they can be easily "healed," with minor damage easily repaired simply by grading and/or recompaction.

* To avoid getting too cumbersome with the terminology, "gravel" is used here to denote roads surfaced with crushed rock, crushed gravel, or pit-run sandy gravel. When it is important, the distinction between the three different materials will be made explicit. "Sealed roads" include those surfaced with asphalt, concrete, or a surface treatment such as chip seal. "Unsealed roads" include roads surfaced with crushed rock, crushed gravel, pit-run sandy gravel, or earth.

Table 14.5 Maintenance treatments for surface-treated pavements

Treatment	*Description*	*Application*
Manual patching	Spreading and compaction of premixed asphaltic materials (hot, cold, or recycled mix), and compaction manually or by machine	Repairs potholes, depressions
Machine patching	Machine spreading and compaction of premixed asphaltic materials (hot, cold, or recycled mix)	Repairs major surface distresses such as potholes, depressions, alligator cracking, rutting, and distortion—note, does not solve the problem if root cause is structural weakening
In situ recycling with manual chip seal	Machine removal of surface defects followed by manual chip seal	Removes distortion
Manual spray patching	Manual application of emulsified binder, followed by spreading of aggregate	Repairs potholes, cracking
Manual chip seal	Manual application of emulsified binder with a manual spray bar, followed by spreading of aggregate with a chip spreader and mechanical compaction	Repairs localized distresses
Machine chip seal	Machine application of emulsified binder followed by machine spreading of aggregate	Repairs localized distresses
Fog seal	Machine application of a slow-setting emulsified binder diluted with water, may be followed by machine spreading of sand	Seals small cracks and surface voids
Manual burn and seal	Elimination of excess surface binder by open-flame burning, sealed with heated aggregate or manual chip seal	Eliminates excess surface binder

Source: Chong, G.J., Phang, W.A., and Wrong, G.A., *Manual for Condition Rating of Surface-Treated Pavements*, SP-021, Ministry of Transportation of Ontario, Toronto, Ontario, Canada, 73pp., 1989 © Queen's Printer for Ontario, 2015. Reproduced with permission.

The main requirement is to maintain the cross fall (camber) to shed water [14.1–14.3]. As much as 6% cross fall has been recommended [14.2] although 4% is more typical. A greater cross fall than the 2% usually recommended for sealed road surfaces (asphalt, surface treated, concrete) is usually specified for the following reasons:

- Unsealed roads have rougher surfaces which impede surface drainage
- Unsealed roads develop ruts between grading operations, which can also impede surface drainage

- The grading operation itself, if poorly performed, can result in a flattening of the road cross section—*specifying* 4% or 5% may actually *result* in only 2% or 3% in reality

To keep gravel and earth-surfaced roads in good condition, more gravel needs to be added where weaknesses appear. It is important to add gravel where there are ruts or weaknesses rather than to redistribute the gravel already on the road by cutting high spots and filling low spots. Such a practice weakens the high spots.

14.6.1 Deterioration

The defects of gravel and earth-surface roads can be classified as surficial or structural. Surficial defects tend to affect road user safety and comfort, whereas structural defects are more a concern of agency costs [14.3]. Surficial defects are caused by climatic conditions, unsuitable material, inappropriate grading, or poor compaction [14.3]. Except for unsuitable material, the defects can usually be removed with careful grading, and compaction if needed. Structural defects are more deep seated and are often caused by overstressing of the base, subbase, and/or subgrade. Before determining a repair strategy, the cause of the structural defect should be determined. It may be a lack of drainage, poor compaction, inappropriate material, or insufficient pavement* layer thicknesses.

Corrugations (washboards) are the displacement of material due to tractions on the road surface, such as braking, accelerating, or climbing grades. The road surface develops parallel ridges at right angles to the direction of travel, with a spacing of approximately 0.5–1 m (18 in. to 3 ft) and depths to 0.15 m (6 in.). Granular materials with particle sizes greater than 5 mm (0.2 in.), low plasticity, and limited fines are susceptible to corrugation. Grading too fast can also cause corrugations to form [14.3].

Potholes affect riding comfort and can cause serious damage to vehicles if left to grow large enough. Those affecting vehicles are 0.25 m (10 in.) or more in diameter, and more than 50 mm (2 in.) deep [14.3]. Roads most vulnerable to potholing have no longitudinal grade and have flat cross sections. Water may lie on the surface. Approaches to bridges, which have much flatter cross sections than gravel roads should have, can be problem areas. The formation of potholes is triggered by the stripping of surface material and infiltration of water into the road structure. Fines are pumped away, and the pavement

* "Pavement" is *any* material placed on the natural subgrade at the site. It is not restricted to asphalt or concrete. Thus, even suitable earth dug from the ditches immediately adjacent to the road and placed on the road constitutes "pavement."

disintegrates [14.3]. If the cross fall and longitudinal gradient are sufficient, water doesn't accumulate, and potholes usually do not form.

Rutting is the formation of deformations in wheel paths. The ruts collect water which softens the pavement materials. Ruts are caused by the following:

- Poor grading of material
- Poor compaction
- Inadequate pavement thickness
- Poor surface drainage
- Excessive fines
- Overloaded vehicles
- Vehicle tracking

Slippery surfaces can be caused, in wet weather, by excessive fines in the wearing surface, or a lack of cross fall [14.3], or of course subfreezing temperatures. In dry weather, they can be caused by the accumulation of loose stones on the road surface [14.3].

Surface scour can be longitudinal or transverse, and is caused by the concentrated flow of water over the surface of the road. It can present a dangerous surface and an uncomfortable ride [14.3].

Soft surfaces may result from the selection of unsuitable material [14.3] or insufficient pavement thickness.

14.6.2 Maintenance treatments

Corrugations can be dealt with in the short term by grading—cutting below the trough of the corrugations and reshaping the road surface. For a more permanent solution, depending on circumstances, a clay binder might need to be blended in to the surfacing material, or a high quality crushed aggregate may need to be imported [14.3]. In the worst cases, it may be necessary to seal the affected stretch of the road.

Potholes require the restoration of the road's cross-sectional shape, especially the cross fall, to prevent water from infiltrating in flat spots [14.3]. Severe potholes may require scarification and reshaping the road. The aprons to bridge decks may need to be sealed if severely potholed.

Rutting solutions depend on the cause. Routine grading can replace gravel lost from the ruts. A thicker road section can be constructed if the pavement design is inadequate. Inappropriate materials can be subexcavated and replaced with more suitable material. Truck loads can be regulated more stringently.

Slippery surfaces can be improved by subexcavating inappropriate materials and replacing them with high quality crushed aggregates.

Surface scour can be ameliorated by directing surface water away from the road surface and improving surface drainage adjacent to the road. In an extreme case, it may be necessary to realign the road.

14.6.3 Gravel road maintenance practices

A wealth of practical experience is available in a PIARC (Permanent International Association of Road Congresses) publication [14.7]. It is important to tailor the maintenance practices to local conditions because they are so variable across the globe. An excellent publication on grading practices in the United States has been produced by one of the US LTAP (Local Technical Assistance Program) centers [14.8]. The publication provides details on the operation of motor graders for light and heavy maintenance operations, drainage, the selection and use of good road surfacing gravel, the use of dust palliatives, and innovations.

14.6.4 Dust suppression

Dust generated on gravel- and earth-surfaced roads can have an effect on the economics, health, and the environment on land adjacent to the road [14.3]. McCrea [14.9] cited in Jones [14.10] mentions a case where the health of sheep was adversely affected because their teeth wore down in the dusty environment next to a gravel-surfaced road. Numerous reports cite reduced crop yield in dust-covered fields near gravel roads. Dust can also be an issue of safety for road users.

Dust is caused by the loss of fines from the road surface [14.3]. The fines are loosened by traffic under the right weather conditions, and with the loss of fines, the road surface becomes more permeable, water intrudes, and the road structure weakens. At the same time, there is a loss of binding material for the larger particles in the road surface. They loosen and are kicked off by traffic, and the road surface fabric unravels.

To control the loss of fines as dust, the most effective to least effective treatments are listed as follows [14.3]:

1. Good construction and maintenance practices, including the selection of appropriate surfacing material
2. Mechanical stabilization—adjusting the particle size distribution of the surface course—to form a hard crust
3. The use of chemical dust suppressants, as an adjunct to the first two treatments

The use of chemical dust suppressants can only ever be a short-term mitigation, and should never be considered the sole treatment [14.3].

Good surfacing gravel reduces or eliminates the dust problem. The characteristics of good surfacing gravel are covered in depth in Chapter 6, so it will not be covered here. On the understanding that they are the third category of dust abatement measures, and can only be considered temporary, dust palliatives are covered in the rest of this section.

Effective dust palliatives have the following characteristics [14.3]:

- They prevent particles from becoming airborne
- They resist traffic wear
- They remain in the pavement—they are not leached out, and don't evaporate
- They resist aging
- They don't damage the environment
- They are applied easily with commonly available equipment
- They are workable and responsive to maintenance treatments
- They are inexpensive

There are a number of classes of dust palliatives [14.3]. Table 14.6 lists them, with comments.

From the Australian perspective, Giummarra [14.3] gives guidance on the selection of dust palliatives. A selection flow chart is presented, along with a map of Australian climatic zones as an input in the selection process.

There are cautions on the use of palliatives [14.3]. Chlorides and lignin sulfonates tend to leach out of the road structure in the rain, reducing their effectiveness, and polluting the adjacent ground. They can also become slippery when wet. Chlorides keep the road surface damp but have no cementing action [14.3]. Calcium and magnesium chlorides are deliquescent—they dissolve in moisture absorbed from the air—whereas sodium chloride is, however, hygroscopic—absorbing moisture from the air. Sodium chloride needs high relative humidity (>70%), so is of little use in arid areas. Calcium chloride and magnesium chloride cannot absorb moisture and dissolve at humidity less than 30%–40%, so are also of little use in very dry climates.

The satisfactory applicability of the various treatments depends on a complex relationship between the soil, the weather and climate, and the suppressant itself (Figure 14.2 [14.3]). In any individual case, trials are recommended to determine the best treatment and the recommended application rate.

14.7 MAINTENANCE EQUIPMENT

Many road authorities have manuals on the details of the selection and operation of the myriad pieces of equipment that can be used for road maintenance. An example is the US Washington State Department of Transportation manual [14.11], which details the use of the following:

- Graders, drags, rakes, underbody blades
- Loaders, backhoes, excavators, Gradalls®

Table 14.6 Road dust suppressants

Treatment	*Description*	*Remarks*
Surfactants	Wetting agents: soaps or detergents	Reduce the surface tension of the moisture in the soil—the increased moisture provides a better binding action in the soil
Chlorides	Salts such as NaCl, $CaCl_2$, and $MgCl_2$	Absorb moisture from the air if the humidity is high enough—increased moisture promotes binding action in the fines
Adhesives	Lignin sulfonates	Organic nonbituminous binders which are by-products of pulp and paper production—they make clay fines more plastic at low water content, increasing the binding action—obviously clay particles must be present in the surface course material for this approach to be successful
Electrochemical stabilizers	Sulfonated petroleum or enzymes	Highly ionic, expelling adsorbed water, decreasing air voids, and increasing density
Petroleum products	Recycled waste motor oil	Causes an agglomeration of soil particles—no longer acceptable environmentally because of the pollution by the oil and the possibility that the oil contains heavy metals
Others		Include microbial binders and polymers—constantly evolving

Source: Giummarra, G. (ed.)., *Unsealed Roads Manual, Guidelines to Good Practice*, ARRB Transport Research Ltd., Vermont South, Victoria, Australia, 2000.

- Rotary traveling mixers, hammer mills, portable pug mills
- Water trucks, towed rollers, self-propelled pneumatic rollers, self-propelled steel-wheel rollers, self-propelled vibratory steel-wheel rollers

In recent years, a modification of the excavator bucket called the chuck blade has come into use in North American LVR construction and maintenance. This is a relatively flat blade 1.8 m (72 in.) or more wide, which is connected to an excavator boom. It provides a useful tool for earthworks including the formation of the road subgrade and ditches, particularly in heavy soils.

	Traffic Volumes, Average Daily Traffic			Subgrade Types [2]			Surface Material Fines Content					Climate			Environmental impacts [5]
							Passing 75 μm sieve								
	Light <100	Medium 100–250	Heavy >250 [7]	Clay	Silt	Granular	<5%	5–10%	10–20%	20–30%	>30%	Rainy	Moderate	Dry: >20days <40%RH	
Surfactants											3	1,3			
Salts: $CaCl_2$, $MgCl_2$											3	1,3		8	
Lignosulphonates			6								3	1			
Petroleum products			6							4	4	3			

Dust suppressant performance: Good / Fair / Poor

Notes:

1. May leach out in heavy rain.
2. Subgrade may mix with surface material and have an impact on the quality of fines.
3. May become slippery in wet weather.
4. Difficult to coat all particles and to prevent "dust pockets."
5. Any product listed may have adverse environmental impacts if used improperly.
6. Hard surface crust can encourage the formation of potholes and break-up under heavy traffic.
7. Products may require greater amounts or more frequent applications under heavy traffic.
8. $CaCl_2$ is likely to behave poorly under these climate conditions.

Figure 14.2 Dust palliative effectiveness. (From Giummarra, G. (ed.)., *Unsealed Roads Manual, Guidelines to Good Practice*, vermont South, Victoria, Australia: ARRB Transport Research Ltd., 2000. Used with permission of ARRB group Ltd.)

REFERENCES

14.1 O'Flaherty, C.A. (ed.). 2002. *Highways: The Location, Design, Construction, and Maintenance of Road Pavements*, 4th edn. Oxford, U.K.: Butterworth-Heinneman, 553pp.

14.2 Lay, M.G. 2009. *Handbook of Road Technology*, 4th edn. New York: Spon Press, 933pp.

14.3 Giummarra, G. (ed.). 2000. *Unsealed Roads Manual, Guidelines to Good Practice.* Vermont South, Victoria, Australia: ARRB Transport Research Ltd.

14.4 Jordan, J.R., Logue, G., and Anderson, G.M. 1980. Review of NAASRA study of road maintenance standards: Costing and management. *Proceedings of 10th ARRB Conference* 10(2):232–249.

14.5 Chong, G.J., Phang, W.A., and Wrong, G.A. 1989. *Manual for Condition Rating of Flexible Pavements*, SP-024. Toronto, Ontario, Canada: Ministry of Transportation of Ontario, 111pp.

14.6 Chong, G.J., Phang, W.A., and Wrong, G.A. 1989. *Manual for Condition Rating of Surface-Treated Pavements*, SP-021. Toronto, Ontario, Canada: Ministry of Transportation of Ontario, 73pp.

14.7 TRL PIARC. 1994. *International Road Maintenance Handbook*, Vol. 2, Maintenance of unpaved roads. Crowthorne, U.K.: The Transport Research Laboratory (TRL).

14.8 Skorseth, K. and Selim, A.A. 2005. Gravel roads, maintenance and design manual. Washington, DC: U.S. Department of Transportation, Federal Highways Administration. South Dakota Local Transportation Assistance Program Report No. LTAP-02–002, April, 2005.

14.9 McCrea, P.R. 1984. An assessment of the effects of road dust on agricultural production systems. Lincoln College, New Zealand: Agricultural Economics Research Unit. Research Report No. 156.

14.10 Jones, D. 2008. Road dust management: State of the practice. In *Proceedings First Road Dust Management Conference*, San Antonio, TX, 15pp.

14.11 Strombom, R.D. 1987. *Maintenance of Aggregate and Earth Roads.* Olympia, WA: Washington State DoT, 71pp.

14.12 AASHTO. 1993. *AASHTO Guide for Design of Pavement Structures.* Washington, DC: American Association of State Highway and Transportation Officials (AASHTO).

14.13 Hajek, J.J., Smith, K.L., Rao, S.P., and Darter, M.I. 2008. *Adaptation and Verification of AASHTO Pavement Design Guide for Ontario Conditions.* Toronto, Ontario, Canada: Ministry of Transportation of Ontario.

14.14 MTO. 1990. *Pavement Design and Rehabilitation Manual.* Toronto, Ontario, Canada: Surveys and Design Office, Highway Engineering Division, Ministry of Transportation of Ontario (MTO). SDO-90–01.

14.15 Kennedy, C.K. and Lister, N.W. 1978. Prediction of pavement performance and the design of overlays, TRRL Report LR833. Crowthorne, U.K.: The Transport Research Laboratory.

Chapter 15

Pavement maintenance and rehabilitation management

15.1 PRIORITY PLANNING AND PROGRAMMING

Whatever the type of pavement or road under consideration, the salient problem is that the budget available for roads will always be insufficient. Therefore, rational methods for controlling expenditure on road construction, maintenance, and rehabilitation are needed. One of these three aspects of road work must not be favored over another. A balance must be struck as the engineer or manager responsible for the road assets attempts to obtain the greatest benefit from the funds that are available.

The problem is one of deciding which projects are funded, when. The "which" is addressed with the planning of priorities in a list of candidate projects, and the "when" is set by the programming of projects. If a completely rational process is followed, the prioritizing and programming processes deal with road construction, maintenance, and rehabilitation on an equal basis.

Various schemes for prioritization, ranging from the simplest to the more sophisticated, can be identified. Candidate projects can be ranked by some appropriate criterion, such as one of the following:

- By road class: all work on first-class road sections is done first, work on second-class road sections is done next if funds are left over, and so on down the road classes until the budget is exhausted
- By use, for example, by traffic volume: priority for candidate sections is assigned in order of decreasing traffic volume
- By cost: maintenance is done on the most expensive road sections first
- By condition, strength, roughness, or some other parameter indicating the health of the road section's pavement

Other schemes prioritize the candidate projects by optimization routines aimed at maximizing overall benefits or minimizing overall costs [15.1].

15.2 COSTS AND BENEFITS

Costs and benefits can be identified for the agency, authority, or party responsible for the road network, the road users, and for the public at large. In some cases, benefits are really decreases in costs attributable to some action.

Costs and benefits of interest to the road owner (agency, authority) include [15.2]

- Road and pavement maintenance costs
- Future rehabilitation costs
- Public attitude and complaints
- Costs related to safety

For road users, costs and benefits revolve around [15.2]

- Travel time
- Vehicle operating costs
- Accident costs
- "Discomfort costs"

Finally, the public in general is concerned with costs and benefits such as [15.2]

- The influence of commercial transportation on the prices of goods, and on employment
- The environment (noise, pollution, vibration)
- Aesthetics
- Accessibility to other transport facilities

Identifying, quantifying, and dealing with these costs and benefits for pavements in the public sector can be a very difficult exercise as the inputs can be very difficult to quantify and may be outside the control of the authority. There can be considerable pressure brought to bear on road authorities to maintain or upgrade particular roads and pavements—there is the temptation to have politics, rather than a rational, systematic approach to the decisions influence the choices.

In the resource industries, companies running their own private road networks can actually have an easier task in coming to grips with the problem. The company is both road network owner and user or has a direct connection to hauling contractors who are themselves the users of the roads. Benefits and costs are much more direct and are far easier to calculate than is the case in the public sector, and the costs incurred in one aspect of the road and transportation work produce direct benefits in another. There is a far greater degree of control available to resource operations than their public counterparts.

15.3 MANAGEMENT AT THE NETWORK AND PROJECT LEVELS

A distinction is made between the decisions taken at the network level and at the project level [15.1].

At the network level, the complete system of pavements is considered. Assessments of the "average" condition of the network can be made, now and into the future. Review of the network situation leads to the proposal of various specific candidate projects and the planning of their priority. From there, management moves to the project level. At this level, the specifics of the engineering of the maintenance or rehabilitation of each particular section of pavement are identified and developed. Assessment is much more specific, and planning is much more detailed. Candidate projects must compete with one another for funding, according to the rationale developed in the management system.

15.4 PAVEMENT OUTPUTS

How does the engineer or manager know how good a pavement's health* is?

Formal pavement management took a leap forward when engineers stopped thinking in terms of "good" and "bad" roads and pavements. This good/bad, black/white concept is not realistic. In reality, a road or pavement begins its life in good health, and then continuously deteriorates. The rate of deterioration may speed up as time passes or with traffic, as shown in Figure 15.1, where pavement condition is shown plotted against time or traffic. Some pavement engineers prefer to say that the road or pavement has a certain design life, and with time or traffic, that life is used up.

A minimum tolerable pavement condition can be specified (assuming it can be measured). As shown in Figure 15.2a, if the pavement reaches this minimum condition, some sort of remedial action should be taken (e.g., grading, resurfacing, or adding significantly more aggregate). If the action is indeed taken, the significant improvement in condition will be exhibited as a jump in the curve, but then the whole process of deterioration will begin again, and the condition curve will decrease toward the minimum tolerable value once again.

Different remedial actions will have a greater or lesser effect on the condition of the pavement and on the condition curve. A simple grading of the road will have much less impact than substantial regraveling, for instance.

* Although to the reader "health" may seem unusual term to use, it will be used to denote the general condition of the pavement because "condition" has a very specific meaning in pavement engineering.

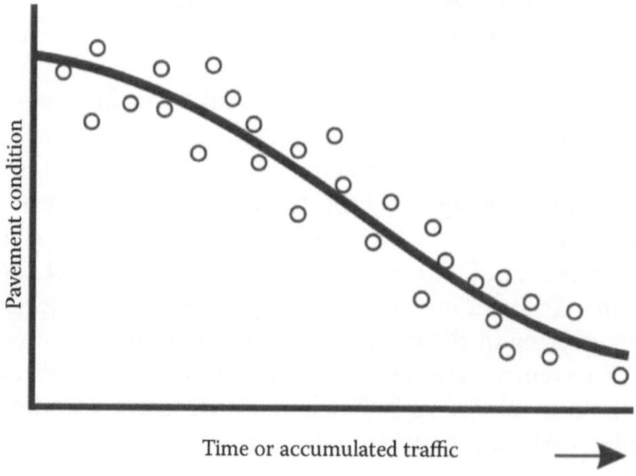

Figure 15.1 Pavement condition data for road sections with similar characteristics and curve-fit "master curve" modeling pavement deterioration.

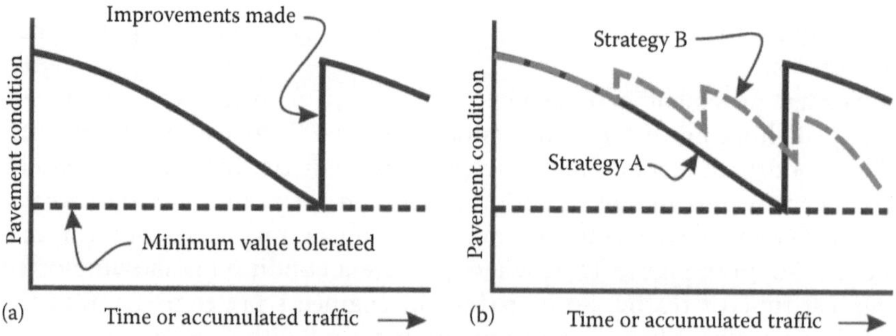

Figure 15.2 Interpreting the pavement deterioration curve: (a) condition deteriorates to a minimum tolerable value, the pavement is treated, and the curve jumps to a high value; (b) two different rehabilitation strategies result in markedly different pavement deterioration curves.

This gives rise to the concept of pavement maintenance and rehabilitation strategies, Figure 15.2b. Strategy A in the figure is evidently a matter of letting the pavement deteriorate over a fairly long time (or with the passage of many vehicles) and then doing some very substantial maintenance or rehabilitation work. On the other hand, it can be seen that Strategy B involves fairly frequent treatments of relatively smaller effect.

The economics of the two strategies could be very different, and it is the task of the pavement management system (PMS) to explore those differences. A comprehensive PMS will encompass both engineering considerations and economic considerations.

15.4.1 Structural capacity, roughness, pavement condition, and safety

There are four important indicators (outputs) of road or pavement health [15.2], which are used to measure how much the pavement has deteriorated or how much of its life has been used up:

1. Structural capacity: the ability to carry the design axle load
2. Riding comfort: vehicle passenger comfort as related to pavement roughness
3. Condition: surface defects—distresses—such as cracking, rutting, and distortion
4. Safety: usually measured in terms of skid resistance but also includes other factors such as pavement light reflectance

Assuming an appropriate measure for each of these outputs can be devised (it can), decisions on the timing of maintenance or rehabilitation can be made on the basis of one or more of the outputs reaching a predetermined maximum or minimum tolerated value. In Figure 15.3, rehabilitation of the pavement was undertaken when the riding comfort reached the minimum tolerated value.

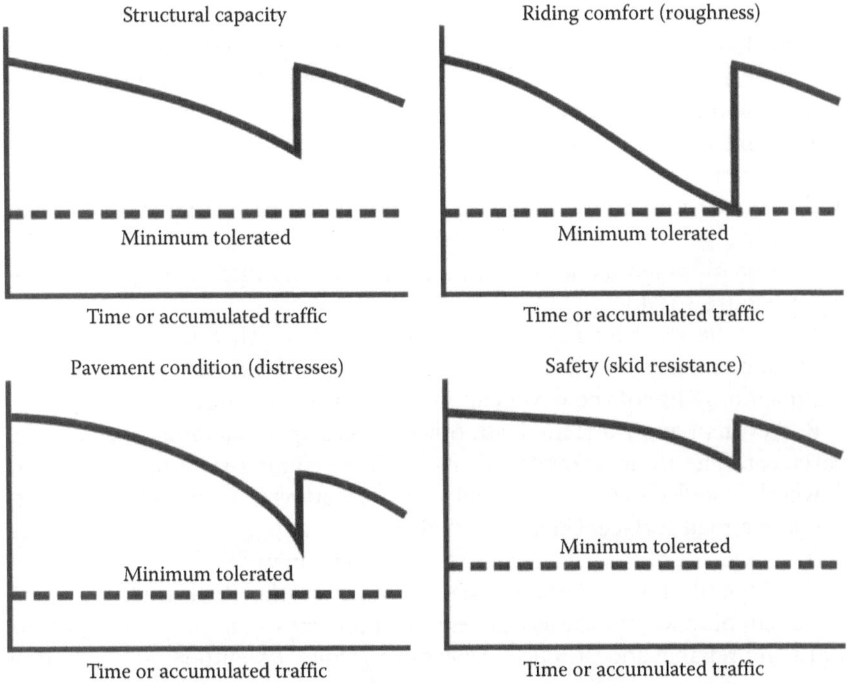

Figure 15.3 Pavement rehabilitation is triggered when one of the four pavement outputs reaches the tolerable limit (riding comfort, in this example).

15.4.2 Measurement of structural capacity

Usually, nondestructive test methods are adopted to measure and record structural capacity. The available equipment may be grouped as follows [15.1]:

- Static deflection devices: for example, plate bearing test, Benkelman beam
- Vibratory devices: for example, Dynaflect
- Impact devices: for example, falling weight deflectometer (FWD)

In the plate bearing test, a static load is imposed on a circular plate, usually 300 or 760 mm (12 or 30 in.) in diameter [15.3]. The deflection of the plate under the known load is then used in one of the many design routines to estimate how many passes of the design axle load it would take to cause the pavement or road to reach some failure criterion (see Chapter 7). That criterion might, for instance, be a maximum rut depth. The tests can be done with a high degree of precision, but they are time consuming and expensive to perform, and result in little data being collected.

On the other hand, a great deal of data is collected in the Benkelman beam test and therefore a much more representative picture is acquired. The beam is essentially a lever supported on three short legs (Figure 15.4a). One end of the beam is inserted between the tires of a dual wheel on a single axle loaded to 80 kN (18,000 lb). By observing a dial gauge bearing on the other end of the beam, fixed to a stationary support, the engineer can determine the rebound of the pavement as the loaded truck is driven ahead to the next station. Stations are usually spaced at 10 m (33 ft) or more.

At a station spacing of 10 m (33 ft), a three-person crew (truck driver, and two carrying and positioning the beam and recording the data) can manually log at least 1 km (0.6 mile) of road in an hour. Higher productivity can be obtained using self-contained, truck-mounted automatic equipment (Figure 15.4b).

Various design routines are available [15.4] which take the pavement deflection and its statistics as inputs together with other factors, to estimate the remaining life of the pavement in design axle loadings.

Rather than apply a static load, other devices apply vibratory loads to the pavement. The trailer-mounted Dynaflect uses counter-rotating unbalanced flywheels installed in the unit to apply a cyclic loading to steel load wheels resting on the road surface. Under the tongue of the trailer, a string of geophones senses the shape of the pavement surface's deflection under the peak loading.

The Dynaflect was perhaps used as much as the Benkelman beam in Canadian practice, for the assessment of structural strength. It is straightforward and reliable and produces repeatable results. Its output can be used in pavement design routines, or, in some jurisdictions, the output is used in correlations with Benkelman beam deflections, and then the correlated deflections are used in the design routines to determine remaining pavement life.

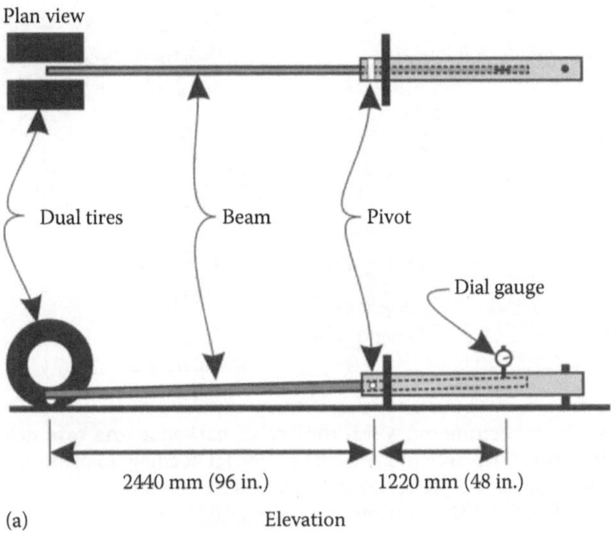

(a)

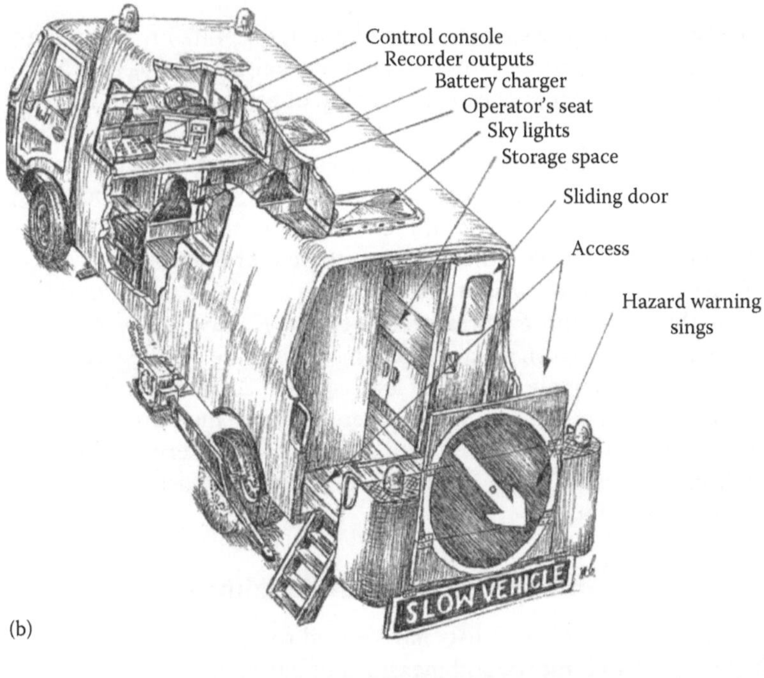

(b)

Figure 15.4 Benkelman beam: (a) manual apparatus; (b) automated, truck-mounted apparatus. (Line drawing copyright Maurice Chapman. Reprinted by kind permission of Linda Cook.)

Figure 15.5 Falling weight deflectometer. With the trailer parked at the test site, the plate is lowered onto pavement surface, impact loading applied, and the instantaneous pavement deflections measured by the string of geophones radiating out from the plate center.

The Dynaflect imposes a peak load of about 4 kN (0.4 tons) on the pavement surface [15.4], appreciably less than a legal truck wheel load, and far less than the wheel loading becoming typical of private resource access roads. For mining haul roads, wheel loads can be as much as 2000 kN (220 tons).

A school of thought is that because pavement structures can behave non-linearly, the loading used to assess the pavement structure should be close to the actual service wheel load, if the test strains in the pavement structure and therefore the surface deflections are to be representative.

The use of the falling weight deflectometer (FWD) is more common now and it ought to give more representative results (Figure 15.5). With a heavy captive weight falling on an anvil resting on a plate on the pavement surface, the FWD can apply an impact load of up to 250 kN (56,000 lb) on the pavement [15.4]. This device also uses geophones to measure the deflections of the pavement surface under this loading and to determine the shape of the deflection bowl. Design routines are available to determine pavement life from these inputs.

15.4.3 Measurement of pavement roughness

Pavement engineering specialists have seized roughness as one of the more easily obtained and understood measures of pavement performance. A survey of the wide variety of methods of estimating pavement roughness is given in Table 15.1.

The riding comfort index (RCI) is a subjective measure of roughness, taken from a panel of raters, who have been driven along the road in a "late model

Table 15.1 Road roughness measuring methods/equipment

Class 1: Precision profiles	Most accurate
Digital incremental profiler ("Dipstick")	
Rod (staff) and level	
Transport Road Research (TRL) profilometer	
Class 2: Other profilometric methods	
Automatic Road Analyzer (ARAN)	
California profilograph	
Dynatest RSP Laser Test System	
K.J. Law Intertial Profilometer	
Longitudinal Profile Analyzer—France	
Road Surface Tester (Laser RST)—Sweden	
Road Tester 3000	
South Dakota Road Profiler	
Class 3: Response-type devices	
K.J. Law Roughness Surveyor	
Mays Ride Metre	
Portable Universal Roughness Device (PURD)	
Walker Roughness Device	
Class 4: Subjective ratings from panels	Least accurate
Riding comfort index, RCI	
Present serviceability index, PSI	

Source: TAC, *Pavement Design and Management*, Transportation Association of Canada (TAC, formerly RTAC), Ottawa, Ontario, Canada, 389pp., 1997.

sedan". The scale in Figure 15.6 is used by the raters to assess the pavement. In Canada, the RCI, a scale from 0 to 10 is used. The U.S. equivalent is the present serviceability index (PSI), which is essentially the same thing, running on a scale from 0 to 5. Clearly this is a subjective way in which to rate pavement roughness, but it has long been used. No longer are panels of raters driven down stretches of road, but the ratings live on in correlations made between the output of various mechanical devices and RCI (or PSI).

At the opposite end of the spectrum are the devices which actually measure the longitudinal road profile, sometimes in minute detail. The "Dipstick" for instance is a handheld instrument [15.5], with a gauge length of 300 mm (12 in.). The other profiling devices are less tedious to use and much more appropriate for road work. What typifies them is that they provide a true representation of the actual profile, from which statistical parameters characterizing roughness can be drawn.

The response-type devices do not involve direct measurement of the road profile. Rather, a vehicle or trailer is driven or towed along the road

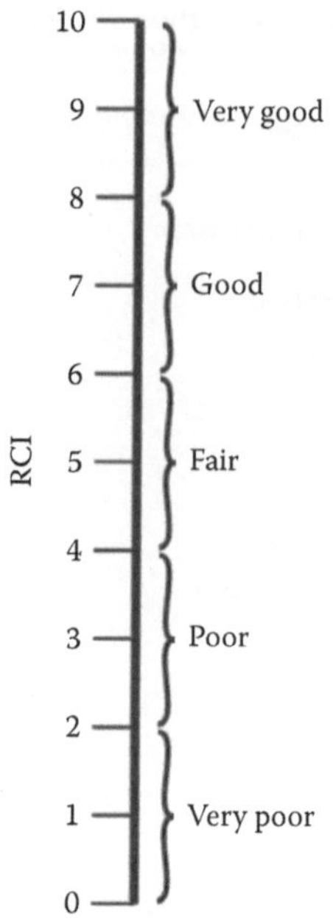

Figure 15.6 Scale for riding comfort index (RCI).

at constant speed, and its suspension's response to the roughness of the pavement recorded. The simplest is perhaps the Mays Ride Metre which has a mechanical device, connected to a car's back axle, which counts the number of times the axle bounces from its neutral position. Given that all response-type devices measure the reaction of a vehicle suspension to pavement roughness, their outputs will depend on the mechanical characteristics of the suspension and the vehicle itself (masses, spring rates, damping coefficients), and the speed at which they are driven.

The mechanical characteristics can be standardized in the quarter car simulation (one wheel equals a "quarter car"), as shown in Figure 15.7. If the masses, spring rates, and damping coefficients meet the standard, and the vehicle is driven at a constant speed of 80 km/h (50 mile/h) along the road section being assessed, then the statistics computed from the response record can be used to determine the international roughness index, IRI. The IRI is given in terms of mm/km (in./mile) of road length tested (Figure 15.8). Shahin [15.5] provides the basis for the calculation of the IRI.

15.4.4 Measurement of distress

The types of surface distress are particular to the pavement surface. The common distresses observed in asphalt-surfaced, surface-treated, and

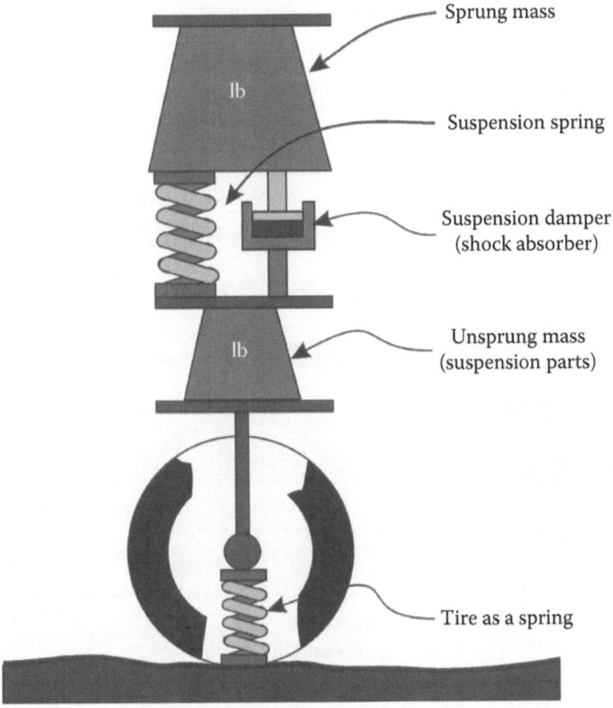

Figure 15.7 Mechanical analogue of the "quarter car."

aggregate-surfaced roads are given in Table 15.2. Distress is expressed in terms of type, severity ("how bad"), and density or extent ("how much"). Each of the distress types has been defined for severity: definitions for ruts, for instance, are shown in Figure 15.9.

A pavement condition index (PCI) is defined. It is an index which runs from 0 (utterly impassable road) to 100 (perfectly defect-free pavement surface), which takes into account all the surface defects observed in the representative sections.

15.4.5 Measurement of skid resistance

Usually the skid resistance of the pavement surface is adopted as the measure of road safety. In addition to a portable pendulum apparatus, suitable only for small areas, two types of skid resistance measuring devices are commonly used:

1. Locked-wheel trailers
2. SCRIM ("sideways force coefficient routine investigation machine")

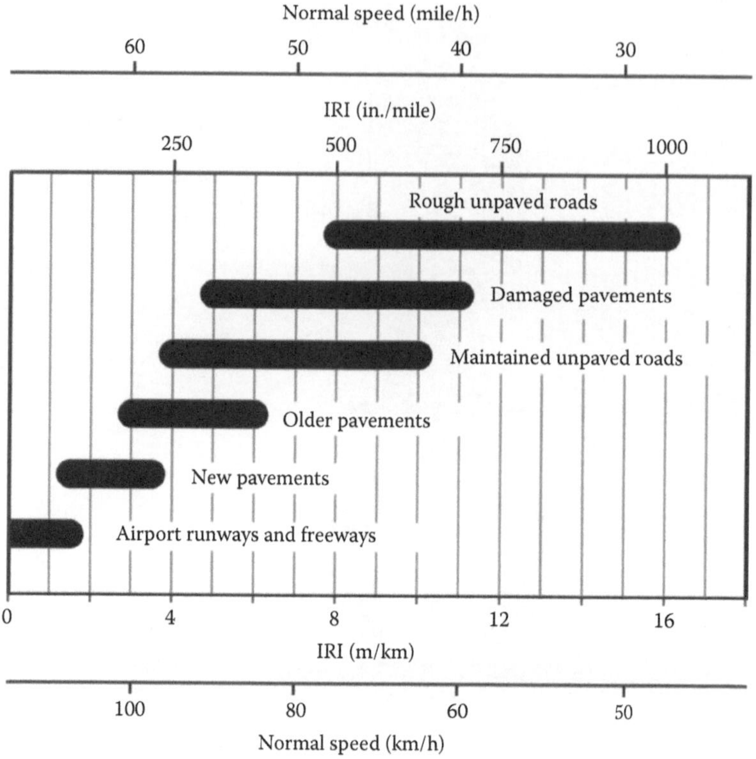

Figure 15.8 International roughness index (IRI) scales. (From TAC, *Pavement Design and Management*, Transportation Association of Canada (TAC, formerly RTAC), Ottawa, Ontario, Canada, 389pp., 1997.)

The locked-wheel trailer apparatus has two wheels, either of which can be locked against rotation as the trailer is pulled along. Load cells in the trailer's hubs monitor the skid resistance offered to the locked wheels. The test is done on wet pavement, with water flushed onto the pavement just ahead of the skidding tires.

The SCRIM test, very common in the United Kingdom, uses a vehicle with an additional wheel which can be angled at a skew to the direction of travel (Figure 15.10). The wheel is free to rotate, but given that its axis is angled to the direction of travel, it exerts an axial force on its bearing. That force can be calibrated for skid resistance. Again, the test is done on wet pavement, with water flushed onto the surface just ahead of the skewed wheel.

15.5 REFERENCING SEGMENTS

It is necessary to break down the complete pavement network into manageable units. It is desirable to work with sections with homogeneous

Table 15.2 Pavement distresses by pavement type

	Asphalt surfaced	*Surface treated*	*Gravel surfaced*
Surface defects	Raveling	Loss of surface cover	Loose gravel
	Flushing	Streaking	Dust
	Potholes	Flushing	Potholes
	Pavement edge breaks	Potholes	Breakup
	Manholes and catch basins	Pavement edge breaks	
Surface deformation	Rippling and shoving	Rippling	Wash boarding
	Wheel track rutting	Wheel track rutting	Wheel track rutting
	Distortion	Distortion	Flat or reverse crown
	Utility trenches		Distortion
Cracking	Longitudinal	Longitudinal	
	Transverse	Transverse	
	Pavement edge	Pavement edge	
	Map	Alligator	
	Alligator		

Sources: Chong, G.J. et al. *Manual for Condition Rating of Surface-Treated Pavements, Distress Manifestations*, SP-021, Ministry of Transportation of Ontario (MTO), Toronto, Ontario, Canada, 73pp., 1989; Chong, G.J. et al. *Flexible Pavement Condition Rating Guidelines for Municipalities*, SP-022, Ministry of Transportation of Ontario (MTO), Toronto, Ontario, Canada, 95pp., 1989; Chong, G.J. and Wrong, G.A., *Manual for Condition Rating of Gravel Surface Roads*, SP-025, Ministry of Transportation of Ontario (MTO), Toronto, Ontario, Canada, 61pp., 1989.

characteristics, ranging from traffic volumes through subgrade soil types. There are four basic methods of dividing and identifying the segments of the pavement network [15.1]:

- Route milepost
- Node link
- Branch section
- Using geographic information system (GIS) functions

In the route-milepost system, routes are identified (for instance, numbered state or provincial highways), an origin defined, and distances along the route measured from the origin. If the node-link system is used, significant points such as intersections or bridges are selected as nodes, and the roads which run between the nodes designated as links. The branch-section system resembles a tree, and the homogeneous parts of each branch are designated as sections. Geographic information system (GIS) approaches make use of mapping coordinate systems and the referencing of specific roads and road sections to that coordinate system.

There are two opposing forces at work, when one is deciding the important matter of which referencing system to use. On the one hand, it should

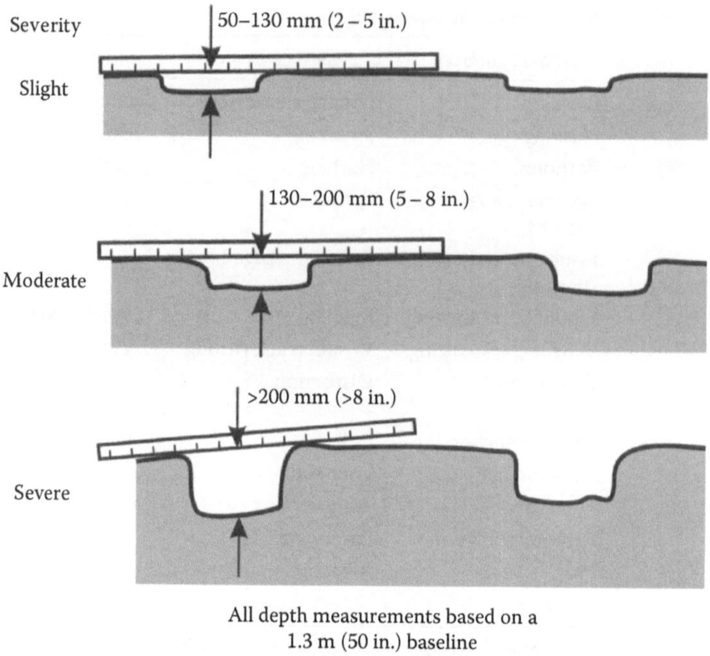

Figure 15.9 Definitions of severities for rutting distress for gravel-surfaced roads. (From Chong, G.J. and Wrong, G.A., *Manual for Condition Rating of Gravel Surface Roads*, SP-025, Ministry of Transportation of Ontario (MTO), Toronto, Ontario, Canada, 61pp., 1989.)

be easy to find specific sections in the field. Systems with fixed, equal length sections facilitate work in the field, but homogeneous sections may or may not result. On the other hand, there are advantages in dividing sections into units with homogeneous characteristics, but their lengths will likely not be equal and they will be more difficult to find in the field. The advantages and disadvantages of the two types of sectioning should be very carefully weighed, as once the issue is settled, large amounts of time and effort will go into the building of the database, and it will be very difficult and expensive to change systems once resources are committed. Developments in GIS such as dynamic segmentation have eased the problem [15.6].

15.6 AN APPROPRIATE PAVEMENT MAINTENANCE AND REHABILITATION MANAGEMENT SYSTEM (PMS) FOR LVR

For LVR, typically very little historical engineering data on the pavement network is available; some LVR authorities may never have had a pavement survey done. It is unlikely that funds will be available to carry out a large,

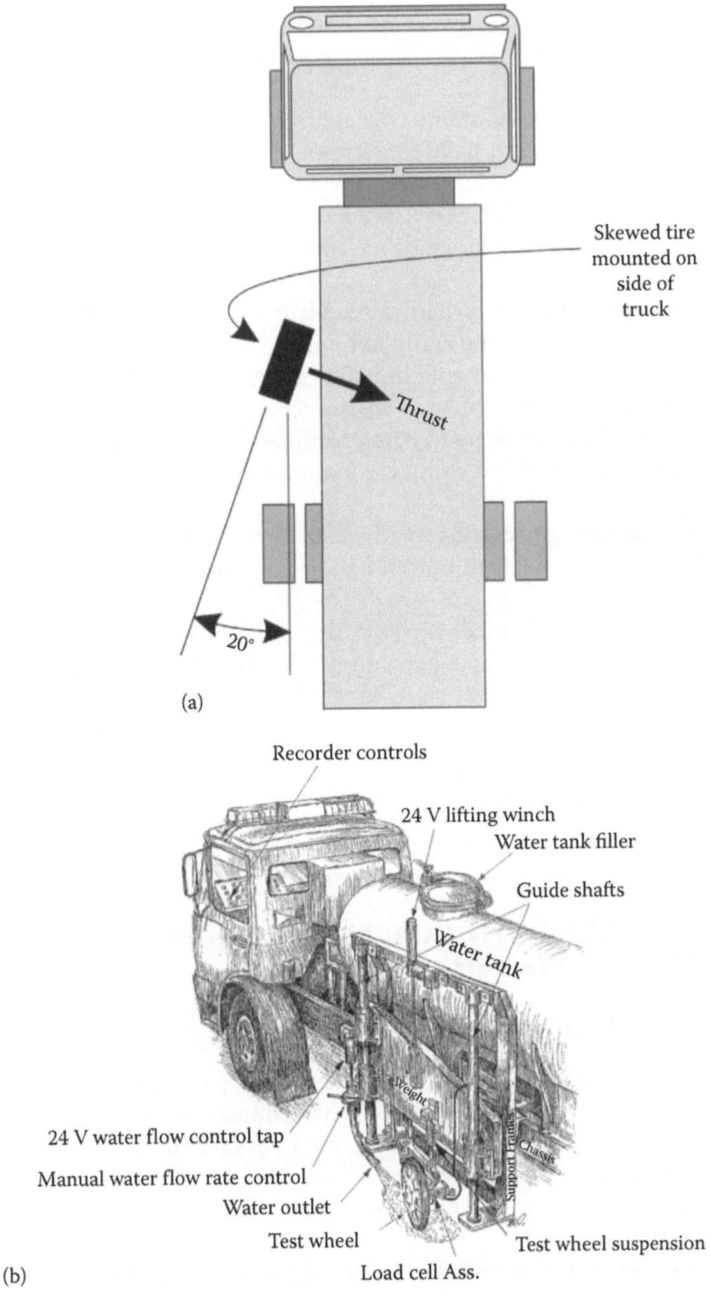

Figure 15.10 SCRIM apparatus for measuring pavement skid resistance: (a) plan view schematic; (b) oblique view. (Line drawing copyright Maurice Chapman. Reprinted by kind permission of Linda Cook.)

complex, or comprehensive data collection program to provide the missing data. An appropriate pavement management system (PMS) may need to be designed on the basis of just one simplified, rapidly collected set of pavement data.

What type of data should be collected?

Pavements are said to have four types of outputs, but the kind of PMS that would be useful for LVR would probably have to rely on just one indicator of pavement health. An argument can be made in the LVR context that it should be pavement condition (distress):

- The assessment of structural capacity requires field equipment that LVR authorities are usually not willing to acquire or hire
- The same applies for roughness measurements
- The measurement of skid resistance varies worldwide from being practically never done to being routine—even if routinely performed, it does not provide a complete picture of pavement health

There is a number of reasons to advocate distress measurement, if the reliance on only one pavement output can be supported:

- It can be performed without the use of sophisticated, expensive equipment—visual, "paper-and-pencil" surveys can be quite effective
- Visual distress surveys are amenable to field data collection with robust tablet computers, simplifying the transfer of field data to the office and reducing the chance of transcription errors
- Some distresses are strongly correlated to structural capacity and roughness, thus providing a more complete view of pavement health

15.7 CAPTURING THE CURRENT CONDITION OF THE PAVEMENT NETWORK

15.7.1 ASTM method

ASTM [15.7] presents a method based on the pavement condition index (PCI), in effect a weighted average of the products of the distress severities and densities (extents—amounts) observed on a section of pavement. It is a way of indicating the pavement's condition with a single index. The method takes no account of the pavement section's structural capacity, roughness, or skid resistance.

The following distresses are considered for asphalt pavements:

- Alligator cracking
- Bleeding
- Block cracking

- Bumps and sags
- Corrugation
- Depression
- Edge cracking
- Joint reflection cracking
- Lane/shoulder drop off
- Longitudinal and transverse cracking
- Patching and utility cut patching
- Polished aggregate
- Potholes
- Railroad crossings
- Rutting
- Shoving
- Slippage cracking
- Swell
- Weathering/raveling

A statistical treatment of the pavement network is applied to identify sample units in which the measurements are taken. The distresses from the list above that are exhibited by the pavement within the sample unit are rated for one of three levels of severity ("low," "medium," "high") where each has specific definitions for each distress type. For example, "medium" rutting ranges from 13 to 25 mm (1/2 to 1 in.) depth. The density (extent, amount) of a distress is usually the percentage of the sample area affected by the distress. An iterative, manual algorithm is then followed to calculate the PCI, a combined rating of the distresses' severities and densities in a single number from 0 (impassable) to 100 (perfect, distress-free condition.)

15.7.2 MTO method

A simpler system to determine a pavement condition rating* (PCR) has been published by the Ministry of Transportation of Ontario (MTO) [15.8–15.10]. The distress types addressed depend upon the pavement surface type (Figures 15.11 through 15.13). Three levels of severity, "slight," "moderate," and "severe," are defined for each individual distress type; density (extent, amount) is estimated to be "intermittent" (<20% of the pavement section), "frequent" (20%–50% of the pavement section), or "extensive" (>50% of the pavement section) for each distress. The ratings for severity and density are made visually—no measurements are made for either.

* "Rating" as distinct from "index" in the PCI, to indicate that a visual approach with little or no measurement is involved.

Riding Condition Rating (RCR) 6.6

Riding Condition Rating (At Posted Speed)

10–8	8–6	6–4	4–2	2–0
Excellent	Good	Fair	Poor	Very Poor
Smooth and Pleasant	Comfortable	Uncomfortable	Very Rough and Bumpy	Dangerous at Posted Speed

Pavement Distress Manifestation			Severity of Distress			Density of Distress %			Distress Manifestation Index (DMI)				
			Slight	Moderate	Severe	Intermittent	Frequent	Extensive					
						<20	20–50	>50					
			1	2	3	1	2	3	W_i	S_i	D_i		$W_i \times (S_i + D_i)$
Surface Defects	Ravelling	1		1			1		3.0	2.0	2.0		12.0
	Flushing	2		1		1			1.5	2.0	0.8		4.1
	Potholes	3							3.0				
	Pavement Edge Breaks	4							0.5				
	Manholes and Catchbasins	5							3.0				
Surface Deformation	Rippling and Shoving	6							1.0				
	Wheel Track Rutting	7	1			1			3.0	0.8	0.8		4.5
	Distortion	8			1	1			3.0	3.5	0.8		12.8
	Utility Trenches	9							3.0				
Cracking	Longitudinal	10		1				1	1.5	2.0	3.5		8.3
	Transverse	11		1				1	1.0	2.0	3.5		5.5
	Pavement Edge	12	1			1			0.5	0.8	0.8		0.8
	Map	13							0.5				
	Alligator	14							3.0				
SCORE			0.75	2	3.5	0.75	2	3.5	27.5	Max	193	Σ	47.9

Pavement Condition Rating (PCR)

DMI = 47.9

RCR = 6.6

PCR = 65

Figure 15.11 Example calculation of PCR for asphalt pavements

Riding Condition Rating (RCR) 5.5

Riding Condition Rating (At Posted Speed)

10–8	8–6	6–4	4–2	2–0
Excellent	Good	Fair	Poor	Very Poor
Smooth and Pleasant	Comfortable	Uncomfortable	Very Rough and Bumpy	Dangerous at Posted Speed

Pavement Distress Manifestation			Severity of Distress			Density of Distress %			Distress Manifestation Index (DMI)				
			Slight	Moderate	Severe	Intermittent	Frequent	Extensive					
						<20	20–50	>50	W_i	S_i	D_i		$W_i \times (S_i + D_i)$
			1	2	3	1	2	3					
Surface Defects	Loss of Cover Aggregate	1		1			1		3.0	2.0	2.0		12.0
	Streaking	2							1.0				
	Flushing	3							2.0				
	Potholes	4		1			1		1.0	2.0	2.0		4.0
	Pavement Edge Break	5			1	1			2.0	3.0	1.0		8.0
Surface Deformation	Rippling	6							2.0				
	Wheel Track Rutting	7	1			1			3.0	1.0	1.0		6.0
	Distortion	8		1			1		3.0	2.0	2.0		12.0
Cracking	Longitudinal	9			1		1		1.0	3.0	2.0		5.0
	Transverse	10							0.5				
	Pavement Edge	11			1			1	1.0	3.0	3.0		6.0
	Alligator	12		1			1		3.0	2.0	2.0		12.0
SCORE			1	2	3	1	2	3	22.5	Max	135	Σ	65.0

Pavement Condition Rating (PCR)

DMI = 5.2

RCR = 5.5

PCR = 40

Figure 15.12 Example calculation of PCR for surface-treated pavements

Riding Condition Rating (RCR) 7.5

Riding Condition Rating (At Posted Speed)

10–8	8–6	6–4	4–2	2–0
Excellent	Good	Fair	Poor	Very Poor
Smooth and Pleasant	Comfortable	Uncomfortable	Very Rough and Bumpy	Dangerous at Posted Speed

Pavement Distress Manifestation			Severity of Distress: Slight	Moderate	Severe	Density of Distress %: Intermittent	Frequent	Extensive	Distress Manifestation Index (DMI): W_i	S_i	D_i		$W_i \times (S_i + D_i)$
						<20	20–50	>50					
			1	2	3	1	2	3					
Surface Defects	Loose Gravel	1		1		1			3.0	2.0	1.0		9.0
	Dust	2							0.5				
	Potholes	3	1			1			3.0	1.0	1.0		6.0
	Breakup	4							1.0				
Surface Deformation	Washboard	5		1			1		3.0	2.0	2.0		12.0
	Rutting	6							3.0				
	Flat/Reverse Crown	7							1.0				
	Distortion	8							0.5				
	SCORE		1	2	3	1	2	3	15.0	Max	90	Σ	27.0

Pavement Condition Rating (PCR)

DMI = 27.0

RCR = 7.5

PCRI = 71

Figure 15.13 Example calculation of PCR for gravel-surfaced roads

While the system relies heavily on the observer's skill, it is much faster (and less expensive to implement) than the ASTM method.

Figures 15.11 through 15.13 illustrate spreadsheets for the calculation of the PCR. Note that the weighting factors can be adjusted. The following equations [15.8–15.10] are relevant:
for all pavement surface types:

$$\mathrm{DMI} = \sum \left[W_i \left(S_i + D_i \right) \right] \tag{15.1}$$

for asphalt-surfaced pavements:

$$\mathrm{PCR} = 100 \times \sqrt{0.1 \times \mathrm{RCR}} \times \frac{\mathrm{DMI}_{\max} - \mathrm{DMI}}{\mathrm{DMI}_{\max}} \times 0.924 + 8.856 \tag{15.2}$$

for surface-treated pavements:

$$\mathrm{PCR} = 12.75 + 90 \times \frac{\mathrm{DMI}_{\max} - \mathrm{DMI}}{\mathrm{DMI}_{\max}} - 5.5 e^{\left[\frac{(9.94 - RCR)}{3.46}\right]} \tag{15.3}$$

for gravel-surfaced roads:

$$\mathrm{RCR} = 100 - \left[\mathrm{DMI} + (10 - \mathrm{RCR}) \right] \tag{15.4}$$

where

- DMI is the damage manifestation index
- DMI_{max} is the maximum damage manifestation index possible
- W_i is a distress weighting factor
- S_i is a distress severity (1, 2, or 3 for "slight," "moderate," or "severe")
- D_i is a distress density (1, 2, or 3 for "intermittent," "frequent," or "extensive")

15.7.3 U.S. Army Corps of Engineers (USACE) method

Eaton [15.11] presents a method tailored to unsurfaced (granular) roads. It is essentially an adaptation of the AASHTO method, calibrated to gravel-surfaced roads.

15.8 MODELING THE FUTURE CONDITION OF THE PAVEMENT NETWORK

Decisions on the priority planning and programming of maintenance and rehabilitation depend on the modeling of expected pavement deterioration. In turn, modeling requires records of past deterioration, under known conditions. It is imperative that periodic measurement of pavement outputs be made and records faithfully kept. Eventually enough data is accumulated to permit the construction of models of road deterioration under known conditions. Figure 15.1 depicts a curve-fit model for pavement deterioration for a number of pavement sections within the road network. It is important that besides all having the same pavement surface type, the pavement sections observed must have relatively uniform characteristics (e.g., subgrade soil, traffic, pavement structure) so that "apples are compared to apples."

With an accurate model of past pavement deterioration in hand, it becomes feasible to predict future deterioration of individual pavements, which in turn facilitates prioritization and programming, the selection of treatments, and the estimation of maintenance or rehabilitation costs. The current condition of the pavement section is compared to what is expected by the deterioration model, and one of several schemes adopted to amend the deterioration curve to match the specific behavior of that particular pavement section (Figure 15.14).

Once the predicted deterioration curve is established, in the budgeting phase, the trial treatment year is selected, an appropriate treatment selected, and the cost of treating the section calculated.

15.9 SELECTING THE MAINTENANCE OR REHABILITATION TREATMENT

The basis for the selection of the treatment can be the projected condition of the pavement at the time it is programmed for maintenance or rehabilitation, as shown in Figure 15.15. Upper and lower limits are set for each type of treatment. For example, "minor rehabilitation" (e.g., milling 50 mm (2 in.) of surface asphalt and replacement with a layer of new asphalt) might be selected for pavements having $55 < \text{PCR} < 65$ at the time the pavement section is programmed for work.

In another approach, a matrix of maintenance and rehabilitation treatments tied to major distresses is drawn up, and the treatment selected based on the current dominant distress.

15.10 PUTTING IT ALL TOGETHER

How are the components of pavement maintenance and rehabilitation management put together into a workable system? How does the system support the budgeting effort?

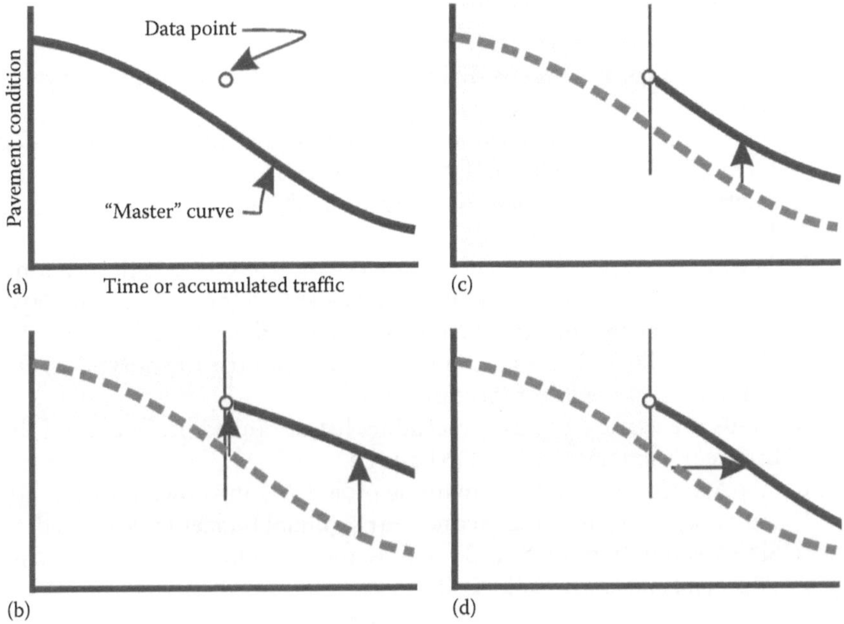

Figure 15.14 Schemes for predicting future pavement deterioration using an adjusted master curve: (a) actual data point does not lie on master curve; (b) master curve adjusted by raising the PCR values in proportion to the time since construction or the last major rehabilitation; (c) master curve adjusted by translating the master curve upward (i.e., adjusting the future PCR values); (d) master curve adjusted by translating the master curve to the right (i.e., resetting the time or traffic values).

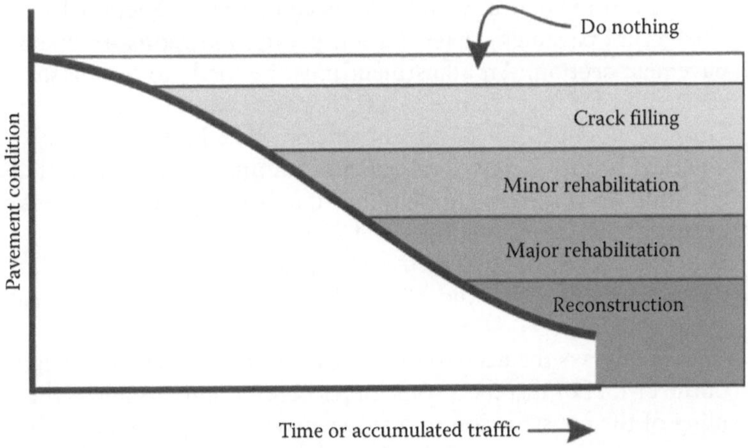

Figure 15.15 Tying the pavement rehabilitation treatment to the pavement condition (PCR value).

Generally, the following steps are taken:

1. The pavement network is divided into sections of uniform characteristics (Section 15.5).
2. Data is collected to determine the current health of each section—for LVR, pavement condition data (distress data) is recommended as the single indicator of pavement health to be determined (Sections 15.6, 15.4.4, and 15.7.2).
3. The full set of pavement sections are ranked according to the criterion adopted (i.e., oldest first, or most expensive first, or most heavily travelled first, or sections in worst condition (Section 15.1).
4. Current PCR for each section is used to assign treatment type and to estimate treatment cost (Section 15.9).
5. Funds are assigned to the descending list of ranked projects until the first year's annual budget is exhausted.
6. Step 5 is repeated for the remaining projects, going down the descending ranked list, until the second year's annual budget is exhausted.
7. Step 5 is repeated for Year 3, Year 4, and so on until the complete list of projects has been addressed.

This is a simplified outline of the steps. There are a number of complications that need to be addressed:

- For those sections not treated until Year 2, it must be remembered that all those pavements will be a year older and will have moved to the right on the pavement deterioration curve. Their PCR values will have dropped somewhat, and the associated treatment may have changed from what it would have been in Year 1 (Section 15.8).
- The actual pavement deterioration curve followed is unique to each pavement section. An adjustment must be made to the master curve to account for this (Figure 15.14).
- Once rehabilitated, a pavement section PCR jumps to a high value, and then begins to deteriorate again, running down the adjusted deterioration curve. Given enough time, it may reappear on the ranked list of sections to be treated (Figure 15.2).
- While the PCR of a treated section jumps to a high value, it may not actually jump all the way to 100—it depends on the treatment (Figure 15.2).
- There is always the need to override the purely systematic assignments of the PMS. For instance, some other consideration, such as the scheduling of the construction of a sewer or water main beneath the pavement may mean the rehabilitation of the pavement should wait a year beyond the year it was originally scheduled. Some facility for moving the timing of rehabilitation forward or backward—while at the same

time tracking the changes in PCR as it moves—is needed. Note that this in turn leaves the PMS vulnerable to political influence in the timing of projects (Section 15.2).

In order to manage all these nuances, it is unlikely that a PMS would be carried out entirely manually for a network of LVR pavements. The data collection can quite easily be accomplished quickly, effectively, and inexpensively manually, as indicated earlier. However, the analysis and budgeting are typically effected with software. Many commercial packages are available.

15.11 MANAGING RESOURCE ACCESS ROADS

The formal pavement management systems used by provincial, state, and municipal agencies have reached a level of maturity, yet the approach has not been embraced by resource industries. At the same time, as has been shown, much of the work would actually be considerably simpler for resource industries to undertake, since they have so much more control over many aspects of road management than public agencies do. The fact that resource industries have not taken up the approach is likely tied to the fact that an appropriate road management system, properly adapted to the resource road design environment, has not yet been devised.

Roughness measurement represents a good option for resource road maintenance and rehabilitation management. A response-type roughness metre which can be mounted in the bed of a pickup truck (ute) or on a light trailer has been devised specifically for use on forest roads [15.12–15.14]. It is suitable for all resource road management.

To be accepted by managers of resource roads, any candidate system must show benefits right from the start [15.15]. Implementation should take place in steps, starting with the setting up of inventories. Merely knowing what the construction and current condition of sections of the road network are would be a great benefit. Periodically updating the records would form the beginnings of datasets needed to produce performance curves. Eventually, the operation keeping faithful records would arrive at the point where it could predict performance and begin to apply economic models. To be able to forecast the cost of various treatments is to be able to minimize costs associated with the maintenance and rehabilitation of the road network.

REFERENCES

15.1 Haas, R., Hudson, R.W., and Zaniewski, J. 1994. *Modern Pavement Management*. Malabar, FL: Krieger Publishing Company, 583pp.

15.2 RTAC. 1977. *Pavement Management Guide*. Ottawa, Ontario, Canada: Roads and Transportation Association of Canada (RTAC).

15.3 Yoder, E.J. and Witczak, M.W. 1975. *Principles of Pavement Design*. Toronto, Ontario, Canada: John Wiley & Sons Inc., 711pp.

15.4 TAC. 1997. *Pavement Design and Management*. Ottawa, Ontario, Canada: Transportation Association of Canada (TAC, formerly RTAC), 389pp.

15.5 Shahin, M.Y. 1994. *Pavement Management for Airports, Roads and Parking Lots*. New York: Chapman & Hall, 450pp.

15.6 Douglas, R.A. and McCormack, R. 1997. Pavement management applied to forest roads. In *Proceedings, 1997 Annual International Meeting*, American Society of Agricultural Engineers (ASAE), Chicago, Paper No. 975031.

15.7 ASTM. 1999. *Standard Practice for Roads and Parking Lots Pavement Condition Index Surveys*. ASTM D 6433-99. West Conshoshoken, PA: American Society for Testing and Materials (ASTM), pp. 736–781.

15.8 Chong, G.J., Phang, W.A., and Wrong, G.A. 1989. *Manual for Condition Rating of Surface-Treated Pavements, Distress Manifestations*, SP-021. Toronto, Ontario, Canada: Ministry of Transportation of Ontario (MTO), 73pp.

15.9 Chong, G.J., Phang, W.A., and Wrong, G.A. 1989. *Flexible Pavement Condition Rating Guidelines for Municipalities*, SP-022. Toronto, Ontario, Canada: Ministry of Transportation of Ontario (MTO), 95pp.

15.10 Chong, G.J. and Wrong, G.A. 1989. *Manual for Condition Rating of Gravel Surface Roads*, SP-025. Toronto, Ontario, Canada: Ministry of Transportation of Ontario (MTO), 61pp.

15.11 Eaton, R.A. 1988. Development of unsurfaced road rating methodology. Special Report 88-5. Hanover, N.H.: US Army Cold Regions Research and Engineering Laboratory (CRREL), 13pp.

15.12 Provencher, Y. and Méthot, F.E. 1994. Controlling the state of the road surface through grading management. Technical Report TR-110. Pointe Claire, Québec, Canada: Forest Engineering Research Institute of Canada (FERIC), 9pp.

15.13 Provencher, Y. 1997. A pavement management system for forestry road networks. In *Proceedings of the International Symposium on Thin Pavements, Surface Treatments and Unbound Roads*. Fredericton, New Brunswick: Geotrans, UNB Natural Resources Geotechnique and Transportation Engineering, Faculty of Forestry and Environmental Management, pp. 265–272.

15.14 FPInnovations. 2015. *Performance Innovation Transport*. Montreal Quebec, Canada: FP Innovations. https://fpinnovations.ca/ResearchProgram/Pages/research-program-pit.aspx, visited January 27, 2015.

15.15 Douglas, R.A. and McCormack, R. 1995. Pavement management systems appropriate to forest haul road networks. In *Proceedings XX World Congress, International Union of Forestry Research Organizations (IUFRO)*, Tampere, Finland, August 6–12, 4pp.

Index